Digital Signal Processing in Audio and Acoustical Engineering

Digital Signal Processing in Audio and Acoustical Engineering

Francis F. Li and Trevor J. Cox

CRC Press
Taylor & Francis Group
Boca Raton London New York

CRC Press is an imprint of the
Taylor & Francis Group, an **informa** business

CRC Press
Taylor & Francis Group
6000 Broken Sound Parkway NW, Suite 300
Boca Raton, FL 33487-2742

First issued in paperback 2023

© 2019 by Taylor & Francis Group, LLC
CRC Press is an imprint of Taylor & Francis Group, an Informa business

No claim to original U.S. Government works

ISBN 13: 978-1-03-265223-8 (pbk)
ISBN 13: 978-1-4665-9388-6 (hbk)
ISBN 13: 978-1-315-11788-1 (ebk)

DOI: 10.1201/9781315117881

Library of Congress Cataloging-in-Publication Data

Names: Li, Francis F., author. | Cox, Trevor J., author.
Title: Digital signal processing in audio and acoustical engineering / Francis F. Li and Trevor J. Cox.
Description: Boca Raton : Taylor & Francis, a CRC title, part of the Taylor & Francis imprint, a member of the Taylor & Francis Group, the academic division of T&F Informa, plc, [2019]
Identifiers: LCCN 2018037679 (print) | LCCN 2018039090 (ebook) | ISBN 9781315117881 (Master ebook) | ISBN 9781466593893 (Adobe Pdf) | ISBN 9781351644150 (ePUB) | ISBN 9781351634663 (Mobipocket) | ISBN 9781466593886 (hardback)
Subjects: LCSH: Sound--Recording and reproducing--Digital techniques. | Signal processing--Digital techniques.
Classification: LCC TK7881.4 (ebook) | LCC TK7881.4.L49 2019 (print) | DDC 621.382/2--dc23
LC record available at https://lccn.loc.gov/2018037679

Visit the Taylor & Francis Web site at
http://www.taylorandfrancis.com

and the CRC Press Web site at
http://www.crcpress.com

Contents

Contents

Preface

Audio or reproduced sound is an important means for people to acquire important and, on many occasions, crucial information from their environment: wake up calls in a hotel room or from an alarm clock at home, morning news from radio or television, public address broadcasts at transportation hubs, media delivered via diverse platforms from a smart phone to a large-screen TV, and numerous telephony conversations or teleconferencing throughout the day at work. It is evident that there is a rapidly growing need for systems and workflows that ensure reliable communication and satisfactory user experience of audio contents over a wide variety of devices and reproduction platforms. Most of the audio contents that users consume these days are reproduced through complicated chains involving many digital processing stages. Digital signal processing (DSP) is at the very heart of modern audio technology. Moreover, to reproduce audio signals, room acoustics need to be taken into account carefully. Room acoustic measurement and equalization in spatial audio reproduction settings involve dedicated signal processing methods.

In the other hand, and with a slightly broader view, acoustical signals of interest are not restricted to those for audible sounds; they extend to sub-sonic and particularly ultrasonic signals, i.e. those with frequencies below and above a typical human audible range. Ultrasonic signals are used for marine underwater wireless communications where electromagnetic (radio) waves do not travel well in salt water due to its electrical conductivity. Sonar is used for underwater object detection. Ultrasonic imaging techniques are used to help medical diagnosis. These rely heavily on signal processing. The principles of physical acoustics in these fields are established; arguably, the technological advancements depend on the development of better transducers and more advanced signal processing methods.

Signal processing rapidly evolved from analogue to digital formats. Traditional DSP has now been developed into a fairly mature area in telecommunications, speech and audio measurement and instrumentation, and many other related disciplines thanks to Shannon's sampling theorem, fast Fourier transform (FFT), and ever increasing computing power. The formation of DSP as a "standalone" discipline or subject is evidenced by a collection of wonderful systematic textbooks published since the 1970s, and DSP being taught and studied as a specialist subject at universities since that time. Traditional DSP centres around quantization, spectrum estimation, and filter design. Essentially, classical DSP is a means of conditioning or manipulating signals. The outputs of such DSP systems are typically a modified version of inputs without changing the nature or type of signals, e.g. noisy speech in and cleaned speech out. Nowadays, more and more feature extraction, pattern recognition, psychoacoustic models, and statistical machine-learning algorithms have come to regular use in audio and acoustical signal processing, opening up a totally modern horizon of DSP in this field. In the context of such modern digital acoustic signal processing, the function of DSP has been extended from signal manipulation to information extraction and decision making, and the system outputs may well take formats other than acoustic signals. For example, in speech-to-text conversion, the speech signal is converted into text, and in speaker recognition, one's voice leads to the decision of who the talker is.

There are dozens of good DSP texts. Some are almost impeccable classics after many editions of evolvements; others have special focal points, distinctive treatment of contents, and intended readers. Before we started preparing for the content for this title, we had asked ourselves again and again: "Do readers in the audio and acoustics society really need another DSP book?" We were soon convinced that a modern version DSP text incorporating machine learning presented from an audio and acoustics angle would be useful to acoustics graduate students, audio engineers, acousticians, and the like to serve as a fast-track learning vehicle to grasp what modern DSP can offer to solve their engineering problems.

Most classical texts on (discrete) signals and systems, filter designs, and relatively modern adaptive filters start from signal and system modelling, and signal statistics, strictly differentiating mathematical models for continuous and discrete signals and systems with some strict mathematical proofs. These are necessary and invaluable training for those in the fields of electronics, communications, and information sciences. Audio engineering and acoustics are allied to these fields, but the main concern is not to develop new DSP methods; rather, the main concern is to apply them. Furthermore, topics similar to signal and system modelling methods have been studied by audio engineers, acousticians, or those converted from other disciplines such as physics and mechanical engineering, in different contexts and discourses. The seemingly similar or identical cores of these can be discussed easily without resorting to detailed mathematical treatment. This book is developed following such thoughts and is intended for the readers who have already received some university training in science and/or engineering.

After summarizing some fundamentals and essentials, the book focuses on DSP in room acoustic measurements, audio related filter design, codecs, spatial audio, and array technologies. It then moves on to adaptive filters, machine learning, signal separation, and audio pattern recognition. The book ends with a chapter discussing DSP in hearing aids.

Most research-informed texts are results of long-term accumulation of research and teaching. This book is no exception. Half of its content is derived from time-tested teaching materials that we delivered in the past 15 years to MSc Acoustics students at Salford. We are indebted to our students for their feedback, and colleagues for their contribution and constructive suggestions.

MATLAB® is a registered trademark of The MathWorks, Inc. For product information, please contact:
The MathWorks, Inc.
3 Apple Hill Drive
Natick, MA 01760-2098 USA
Tel: 508-647-7000
Fax: 508-647-7001
E-mail: info@mathworks.com
Web: www.mathworks.com

About the Authors

Francis F. Li embarked on an academic career in 1986 at Shanghai University. He completed his PhD in the UK and was appointed senior lecturer at Manchester Metropolitan University in 2001, he then joined the University of Salford in 2006. Dr. Li has acquired a broad spectrum of expertise and research interests including audio engineering, signal processing, artificial intelligence, soft-computing, data and voice communications, machine audition, speech technology, architectural and building acoustics, DSP applied to biomedical engineering, novel computer architecture, and control theory. He has undertaken council, charity, government agent and industry funded research projects, and has also published numerous research papers. He has been a committee member and reviewer for numerous international journals and conferences. Dr. Li used to be associate editor-in-chief for the CSC Journal *Signal Processing*, and is currently an associate technical editor for the *Journal of the Audio Engineering Society*.

Trevor J. Cox is a professor of Acoustical Engineering at the University of Salford and a past president of the Institute of Acoustics. He uses blind signal processing methods to model the human response to sounds and to measure acoustic systems. He has presented numerous science documentaries on BBC Radio. He is the author of Sonic Wonderland for which he won an ASA science writing award. His latest popular science book is *Now You're Talking*.

Source: University
of Salford

1 Acoustic Signals and Audio Systems

Sound is waves that propagate through a medium and are perceived by a sensory system such as human ears or a microphone. Acoustics is the science of sound, while audio is often referred to as technically reproduced sound. When an acoustic sensor, for example a microphone, is used to pick up the sound in the air, the change of air pressure due to the wave fronts is converted into a varying potential difference, i.e. voltage over time, which represents the sound. This voltage is, thus, analogue to the acoustic signal of the sound. Such signals are, therefore, called analogue signals. As the sound pressure changes continuously over time, the associated acoustic signal naturally is a continuous function of time, i.e. an analogue signal, which we denote as $x(t)$, $t \in R$.

Audio is about the reproduction of sound. In an audio reproduction chain, sound is first converted to acoustic signals, followed by some processing, storage, or transmission if necessary, and then reproduced, i.e. converted back to sound using an acoustic transducer such as a loudspeaker. The processing stage is at the very heart of an audio system, has always been important, and is getting more and more sophisticated in modern digital audio. Furthermore, acoustic signals and recorded soundtracks are information rich. Speech, event sounds, and music, just to name a few, carry much information of interest. Signal processing finds its applications in the extraction of the information of interest from audio signals, making decisions or judgements. This represents a branch of modern, and an extended area of, acoustical signal processing.

Signals and systems are the major concerns and the foci of study in acoustic and audio signal processing. There are a few dozen excellent textbooks on signals and systems. This book is not an alternative to these texts, which offer detailed systematic treatment of classical signal processing principles and theories, but rather is a specialist text on acoustic and audio signal processing with an emphasis on how signal processing is applied in audio and acoustics, and the recent advancements in this field. This first chapter will present some essential concepts and definitions about signals and systems to pave the way towards the subsequent chapters. It is not an extensive and in-depth summary of a well-established subject for which readers can find more details, backgrounds, and proofs in classical text books.

1.1 SIGNALS AND SYSTEMS

Signals are information carriers. A speech signal delivers semantic meanings, while traffic signals give instructions or directions. Signals are often represented by the changes of a physical quantity and vice versa; for example, a sound signal is given by the change of air pressure due to a vibrating source. Signals are often converted into potential differences or voltages for the ease of manipulation and transmission based

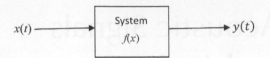

FIGURE 1.1   A block diagram illustration of a system

on the currently prevalent technologies. This is achieved by a sensor. In acoustics and audio engineering, the sensor is more often referred to as a "transducer"; it is one of the two kinds of transducers. (In acoustics, a device that converts sound into electrical signals, such as a microphone, and a device that coverts electrical signal or power into vibration or sound, such as a loudspeaker, are both transducers.) A microphone that converts pressure changes into changes of voltage is an example. Mathematically and in a broad sense, a signal is a variable of one or more dependent variables. The number of dependent variables defines the dimensionality of the signal. Most of the acoustic and audio signals are sound pressure or related voltage over time and, therefore, are one-dimensional signals. Such a signal can be written as $x(t)$, $t \in R$ as we mentioned before.

When we study an object, we often wish to define a boundary between the object under investigation and its environment. The object with a clearly defined boundary is a system. In the context of signal processing, as we are interested in how signals are processed, or in other words, how we manipulate signals, a system is thus often defined as a unit that manipulates the signals, and its environments are the inputs and outputs of the signal processing system. Sometimes, external interferences are present; these are deemed as environments as well.

The input of the system is $x(t)$ and the output after manipulation becomes $y(t)$ and, thus, the system may be viewed as a function $f(x)$ that maps $x(t)$ onto $y(t)$, as illustrated in Figure 1.1.

On rare occasions, an audio system or sub-system can have no input but generates a specific output. A typical example is an oscillator or signal generator. The signals and system shown in Figure 1.1 take the simplest form: there is only one input and one output. It is said to be a single-input-single-output system, or a SISO system. A system may take a multiple-input-single-output (MISO) form, and the other two are multiple-input-multiple-output (MIMO) and single-input-multiple-output (SIMO) systems. In this chapter, we focus on SISO systems. MIMO systems will be discussed in later chapters.

1.2   TYPES OF SYSTEMS BY PROPERTIES

According to the general properties or behaviours of systems, they can be classified from various perspectives.

1. **Linearity**

    A simple and common way to define the linearity of a system is the use of the superposition theorem. If the superposition theorem applies, then the system is linear; otherwise, it is non-linear. More specifically, for a linear system, if two valid inputs $x_1(t)$ and $x_2(t)$ result in system outputs $y_1(t)$ and $y_2(t)$, then the input $ax_1(t) + bx_2(t)$ should result in $ay_1(t) + by_2(t)$, where a and b are any arbitrary real scalars.

Although strictly linear systems are sought after for many acoustic and audio applications, especially for measurements and amplification, non-linear systems are commonplace in modern applications, especially in lossy compression and psychoacoustic codecs. This is because the human auditory system is intrinsically non-linear. It is also worth noting in physical reality, mathematically strict linearity rarely exists, and weak non-linearity is commonplace. For example, a linear amplify will have more or less some no-linear distortions. In engineering practices, such weak non-linearity is often neglected, and we analyse the system as though it was linear.

2. **Time Variance**

Time variance refers to the dependency of a system property on time. In a time invariant system, the input-output relationship is not dependent on time. In such a system, if an input signal $x(t)$ results in a system output $y(t)$, then any time shift t' to the input, $x(t + t')$, results in the same time shift in the output $y(t + t')$. So, one other way to define a time invariant system is that the system function (mapping relation) is *not* a function of time.

Many systems are approximated and modelled as linear and time invariant ones; we refer to them as LTI systems. For the majority of applications in acoustics and audio, time variance is generally a negative attribute. Fortunately for most of the systems, time variance tends to be small in the duration of analysis and operation, so the time invariance assumption is generally valid. However, we will see later there are occasions where time variance may have a significantly negative impact. For example, jitter due to the time variance of a system clock can cause distortions in digital audio, time variance of the room transfer function can affect acoustics measurements, and time variance in telecommunications transmission channels can degrade audio streamed over the network.

3. **Stability**

Stability is an important property of a system, but it is relatively difficult to define and to prove. The most common definition for a stable system is that the system yields a finite output to any valid finite input, though other stability definitions are also used in diverse contexts. More specifically, if the input $x(t)$ satisfies

$$|x(t)| \leq a < \infty \qquad (1.1)$$

then

$$|y(t)| \leq b < \infty \qquad (1.2)$$

where a and b are finite positive real numbers. This type of stability definition is known as a bounded-input-bounded-output stability, or BIBO stability.

4. **Causality**

A system is causal or non-anticipative if its output is only determined by the past and current inputs. A system is said to be non-causal if the output of the system is dependent on inputs in the future. Real-world physical systems are causal. It is also evident that for real-time processing of signals, causality is a requirement.

5. **Memory**

A system is said to have memory if the output from the system is dependent on the past inputs to the system. A system is said to be memoryless if the output is only dependent on the current input. Memoryless systems are the easiest to work with since simple algebraic equations are adequate to describe their behaviours, but systems with memory are more common in signal processing applications. A memory system is also called a dynamic system, whereas a memoryless system is also called a static system. Memory stems from energy storage in the system. Mechanical systems with mass and springs, electronic systems with capacitors and inductors, and digital systems with delay units all have memories. Differential equations for analogue systems or difference equations for digital systems are needed to describe the dynamic behaviours of such systems.

6. **Invertibility**

A system is invertible if the input signal can be determined completely from the output signal. An invertible system has the advantage of being able to determine the input from its output. For many measurement systems, invertibility is vitally important, since the original physical quantity to be measured needs to be recovered from the signals. If F represents the mapping relation from system input to its output, for an invertible system there exists an inverse mapping relation F^{-1} so that

$$x(t) = F^{-1}[y(t)] \tag{1.3}$$

Many LTI systems are invertible, as we shall see later, especially minimum phase systems. However, there are many other systems that are non-invertible. One example is the room transfer function or impulse responses in the time domain. This means it is theoretically impossible to completely remove the room acoustics effect on received sound; nonetheless, various estimations may be used to estimate the original source signals.

1.3 TYPES OF SIGNALS

Signals are probably more complex than systems. There are numerous diverse types of signals, since every single changing physical quantity is giving off some sort of signal, and signals are often stochastic. Signals may be grossly categorised into deterministic and random signals depending upon their statistical features.

1.3.1 DETERMINISTIC SIGNALS

A deterministic signal, as the term itself states, is determined completely at any specific time. In other words, for a deterministic signal, we know exactly what the value is at any given time: in the past, at the present, and in the future. Thus, a deterministic signal can be modelled mathematically as a function of time. In acoustic and audio systems, various deterministic signals are used as probe stimuli to take measurements, calibrate systems, or act as references. A sine wave generated by a signal generator is a typical example. Maximum length sequences and exponential sine sweeps are other commonly used deterministic signals for measurement purposes, and will be discussed in Chapter 3. Some common deterministic signals are summarised below.

1. **Sinusoids**

   Sinusoidal signals are probably the most commonly quoted periodic signals in the field of signal processing. They represent pure tones acoustically. A sinusoidal signal $x(t)$ is given as a function of time t by

   $$x(t) = A\cos(\omega t + \varphi) \tag{1.4}$$

   where A is the amplitude, ω is the radian (or angular) frequency in radians/s and φ is the phase.

   If these parameters are known, the value of $x(t)$ at any past or future times are completely determined. Taking into account the relations between period T in seconds, frequency f in Hertz, and radian frequency in radians/s,

   $$\omega = \frac{2\pi}{T} = 2\pi f \tag{1.5}$$

   We can re-write a sine wave as

   $$x(t) = A\cos(2\pi f + \varphi) \tag{1.6}$$

   This is illustrated in Figure 1.2.

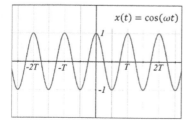
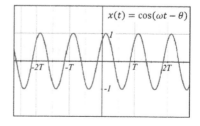

FIGURE 1.2 Cosine waves

2. **Complex sinusoids**
   A complex sinusoid of time is defined by

$$x(t) = Ae^{j(\omega t + \varphi)} \tag{1.7}$$

   A complex sinusoid is not a real-world one-dimensional signal but somehow man-made mathematically. Taking into account the Euler relation

$$e^{j\theta} = \cos\theta + j\sin\theta \tag{1.8}$$

$$x(t) = Ae^{j(\omega t + \varphi)} = A\cos(\omega t + \varphi) + jA\sin(\omega t + \varphi) \tag{1.9}$$

It is evident if the real part (or imaginary part) is taken, the complex sinusoidal signal becomes a sinusoidal one. Using the exponential format of the complex sinusoid, some calculations, e.g. multiplication, can be made more easily. That is one of the motivations of expressing a real signal in a complex form.

3. **Exponential signals**
   Exponential signals reflect many natural processes. Charging a capacitor in a circuit and sound pressure decay after an excitation in a room both follow exponential functions. An exponential signal be written as

$$x(t) = e^{\tau t} \tag{1.10}$$

When $\tau < 0$, there is exponential decay; when $\tau > 0$, there is exponential increase, as shown in Figure 1.3.

4. **Damped sinusoids**
   Many natural oscillations will eventually diminish due to damping; a mass-spring system in the real world is an example. A damped sinusoid can be view as exponential decay enveloped by a sinusoid

$$x(t) = e^{\tau t}\cos(\omega t + \varphi) \tag{1.11}$$

This is illustrated in Figure 1.4.

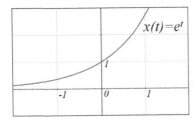 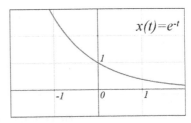

FIGURE 1.3   Exponentials

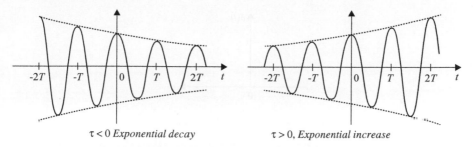

$\tau < 0$ *Exponential decay* $\tau > 0$, *Exponential increase*

FIGURE 1.4 Exponential decay and increase

Damped oscillations reflect many natural processes. We shall see in the next few chapters that exponential decay is associated with system stability and algorithm convergence.

1.3.2 Some Special Testing Signals

There are a number of special testing signals that are used to perform system measurements. These probe stimuli are often idealised mathematical models; some of them can only be approximated in reality but not really implemented precisely.

1. **Unit step function**

   Unit step function $u(t)$ (also known as Oliver Heaviside's step function, shown in Figure 1.5) is a signal used to test certain systems. The unit function was originally developed for solving differential equations. It switches a signal on at the time $t = 0$ and keeps it on forever. The combined use of time-shifted unit steps can be used as a switch to select signals in a specific time interval.

$$u(t) = \begin{cases} 1, & t \geq 0 \\ 0, & t < 0 \end{cases} \tag{1.12}$$

   For example, Equation 1.13 implements a signal selection switch $s(t)$ that switches a continuous signal on at $t = 2$ and then switches it off at $t = 4$.

$$s(t) = u(t-2) - u(t-4) \tag{1.13}$$

2. **Delta function**

   In Dirac's delta function, or simply δ function, $\delta(t)$ is an impulse signal used to acquire the impulse response of a system. It is defined jointly by

$$\int_{-\infty}^{\infty} \delta(t)dt = 1, \tag{1.14}$$

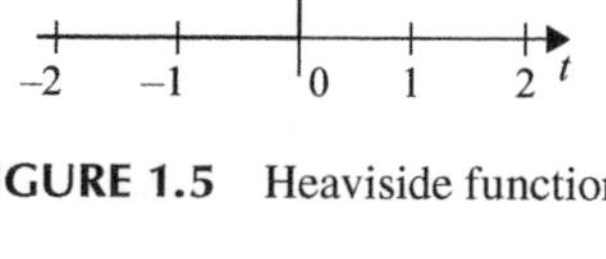

FIGURE 1.5 Heaviside function

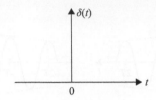

FIGURE 1.6   Dirac delta function

and

$$\delta(t) = 0, \quad \text{when } t \neq 0$$

Figure 1.6 illustrates the δ function. The values of the function are zero every moment in time except at $t = 0$ and the integral over the entire time is 1. Conceptually, it can be viewed as a signal with an infinitely short duration and infinitely high peak that gives unit energy. In a strict sense, it is not a function, since its value at $t = 0$ is not specified. However, this can be viewed as a "generalised function."

In control systems, electronics, and audio, Dirac's delta function is often used as a stimulus to obtain the impulse response of a system, which describes the input-output relationship of the system. Since delta function has infinitely high peak at $t = 0$, but zero value at all other times, physically it cannot be implemented. It cannot be generated electronically either, as electronic circuits have limited outputs. Some compromises have to be accepted and approximation made. In acoustics measurement, a starting pistol or a spark generator, a popping balloon, etc., are used to generate the impulsive excitations for impulse response measurements. For electronically generally pseudo-impulses, a very narrow pulse with a high peak is sought to approximate the impulse. In such a case, the shape of the pulse does not really matter, but the highlight and width matter.

There is a discrete counterpart of the Dirac delta function known as Kronecker delta function.

$$\delta(n) = \begin{cases} 1, & n = 0 \\ 0, & n \neq 0 \end{cases} \quad n \in N \tag{1.14$'$}$$

It will be introduced in the next chapter.

3. **Ramp function**

A ramp function $r(t)$, as illustrated in Figure 1.7, is defined by

$$r(t) = \begin{cases} x, & \text{when } x \geq 0 \\ 0, & \text{when } x < 0 \end{cases} \tag{1.15}$$

It is evident that the derivative of the ramp function is the step function, and the derivative of the unit step function gives the Dirac delta function.

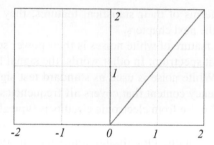

FIGURE 1.7 A ramp function

1.3.3 RANDOM SIGNALS

A random signal is a stochastic process. Many real-world signals, such as environmental noise signals, and speech and music signals, are random or stochastic. Many aspects of acoustic and audio signal processing are indeed statistical processing of random signals.

Random signals *per se* are non-deterministic. Their values are random variables, i.e. signals take random values at various times and, therefore, cannot be modelled deterministically using math functions. They can, however, be characterised statistically.

Random signals can be classified further into stationary and non-stationary signals, depending upon the statistical features. Generally speaking, a random signal is said to be stationary if the statistical features of the signal is time independent. Non-stationary signals mean that statistics of those signals change over time. Now the question is what and how many statistical features we look at; this, in turn, leads to how strict the stationarity is. Strict stationary signals need to satisfy the condition that *all* ensemble statistics are independent of time. Many real-world random signals do not satisfy such a strict condition. We often loosen the requirements slightly and define the so-called weakly stationary signals. Most commonly, we check up to the second order statistics for stationarity. If the mean, variance, and auto-correlation of a signal are independent of time, the signal is deemed as (weakly) stationary. Since the conditions for weak stationarity are relatively easy to satisfy, and weakly stationary signals are the sort of "stationary signals" we often deal with, many authors simply drop the word "weakly." When they refer to the true stationary signals, they use the terminology "strict stationary." We will follow the trend in this book. Furthermore, even though we have loosened the condition for stationarity, many signals are still not "stationary" over a long period of time, e.g. music and speech signals. We often truncate the signals into many short-windowed segments called analytical frames so that signals within these frames can become weakly stationary.

1. **White noises**

   A white noise signal $w(t)$ is a random process in which the value $w(t)$ at any given time t is statistically independent of any values before t. This is a strong definition. Again, there are weaker ones. A slightly weaker definition requires independence between the values of any pair of values $w(t_1)$ and $w(t_2)$ ($t_1 \neq t_2$). An even weaker definition only requires that any pair of values are uncorrelated. There are subtle further details to specify for these white noises in terms of their statistical features; they will be discussed later in this and the next chapters.

One common feature of white noises is their power spectral density is a constant, i.e. a flat spectrum. In other words, the signal has equal power at all frequencies. White noise is used as standard test signals due to its flat broadband frequency content that covers all frequencies equally in power. Residual thermal noise from electronic circuits is typically white noise and perceived subjectively as a hissing noise. It is worth noting that any programmed routine running on a digital computer cannot generate strongly white noises. One can only get pseudorandom or weakly white noises from a digital computer.

2. **Pink noise**

A pink noise is a stochastic process whose spectral density is inversely proportional to the frequency of the signal itself. Compared with white noise, pink noise has more energy (or power) in lower frequencies than higher frequencies. A pink noise shows a 3dB/octave decay rate in its spectrum.

Pink noise may be generated by low-pass filtering a white noise so that the lower end of the spectrum has more energy than the higher end does. The main reason for using a pink noise is because its spectra resemble some real-life sounds and noises. In some measurements, this further offers a more evenly spanned signal-to-noise ratio in various frequency bands.

3. **Other noises named following colours**

There are some other colours used to describe noises. Brown noise has its spectral density inversely proportional to f^2 (frequency squared). Its power spectrum shows a decay of 6 dB per octave. A violet noise is a random noise having a 6 dB increase per octave in its power spectrum. A grey noise is a random noise whose spectrum is shaped according to an equal loudness contour. Therefore, grey noise gives a subjectively flat frequency response.

4. **Speech and music signals**

Speech and music signals are the most processed ones in audio application. Both are non-stationary random signals but can be viewed as piecewise stationary in short observation duration.

Speech is uttered by forcing air from the lungs through the vocal cords and along the vocal tracts. The whole vocal tract of an adult male is about 17 cm long. It starts from the opening in the vocal cords (glottis) and ends at the mouth. It modulates the sound source from the vocal cord source and introduces short-term (around 1 ms) correlations. This can be viewed as a filter with a number of resonances called formants. The frequencies of these formants are controlled by varying the shape of the voice tract, for example by moving the position of the tongue. An established way to model the speech signal is to mimic the vocal tract by short-term filters. As the shape of the vocal tract varies relatively slowly, such filters remain unchanged in short periods of time. As a result, in a 20 to 40 ms time period, the signal can be viewed as stationary.

The production of music signals arguably is simpler than that of speech. However, there are various kinds of musical instruments, with each kind having its own sound production mechanism. The way music is composed also indicates that the signals can be approximated as stationary ones in short periods of time, similar to those of speech.

1.4   STATISTICS OF RANDOM SIGNALS

A signal cannot convey information without some randomness; if the next state or value of a signal is completely predictable and determined from previous values, then it does not need to be delivered! From this point of view, unpredictable future states are the key for information delivery. Information theory uses statistics to deal with the non-predictable states by probability; in other words, statistical estimation.

The values of a random signal are random variables over time. Random signals cannot be modelled mathematically by a function. The characteristics of such signals are often described using signal statistics. Mean, variance, auto-correlation, etc. are all familiar statistical measures used to describe one or some aspects of the signals. Random signals are characterised by statistical parameters.

When we apply statistical analysis to signals, various averaging methods are often considered. There are two general types of statistical averaging, namely time averaging and ensemble averaging. It is important to differentiate these as the statistical meanings are different. Time averaging is an average of the same signal over time, while ensemble averaging is an average for a good number of samples of the kind of signals.

Two most commonly used parameters for random signal characterisation are probability density function and auto-correlation function. The former characterises the variations in amplitude and the latter indicates how a signal changes over time statistically.

1.4.1   Probability Density Function and Moments

Signals found in real-world systems for engineering applications are generally well-behaved. For such signals, two parameters, probability density function and auto-correlation function are sufficient to describe their stochastic behaviour. There are three other parameters that are commonly used, namely mean value, mean square value and power spectral density function. The mean value and the mean square value are important parameters, which give the "average" of the signals and can be derived from the probability distribution. The power spectral density function is related to the auto-correlation function via a Fourier transform.

When we are processing our signals with linear systems, we often design the processing or analyse the results by considering only the first and second moments of the process, namely the following functions: mean auto-correlation and auto covariance. The first and second moments in many cases are sufficient to completely characterise the random process. Strictly speaking, this only happens to a Gaussian process!

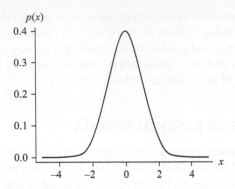

FIGURE 1.8   PDF of normal (Gaussian) distributions as an example

1. **Probability density function and statistics of amplitudes**

   One important property of a random signal is the statistical distribution of its amplitude. Probability density function (PDF) is used to describe such signal amplitude distribution. We denote a signal as $x(t)$, $t \in R$, and if Y represents an observation of $x(t)$, P is the probability, and $p(x)$ is the PDF of $x(t)$, then the PDF is defined such that

$$P(a < Y < b) = \int_a^b p(x)dx \qquad (1.16)$$

   where a and b are real numbers giving bounds of the signal amplitude and dx is an infinitesimally small interval. Direct definition of the PDF is slightly difficult to understand and less clear, so we adopt this indirect definition here.

   It is obvious for all bounded signals (signals having finite amplitude)

$$\int_{-\infty}^{\infty} p(x)dx = 1 \qquad (1.17)$$

   Conceptually, the PDF can be viewed as a function that gives the relationships between the probability of occurrence and the signal values (amplitudes) as illustrated in Figure 1.8.

2. **Cumulative density function**

   Cumulative density function (CDF) is another amplitude-related signal statistical parameter. The CDF $F(x)$ gives the probability that the signal has amplitude below a specific value, say x, so that

$$F(x) = P(-\infty < Y < x) = \int_{-\infty}^{x} p(\xi)d\xi \qquad (1.18)$$

   where ξ is a dummy variable for integration. From the CDF curve, the probability that a signal has a value lower than x is the area under the CDF curve to the left of x.

It is apparent that the CDF and CPD have the following relationship:

$$p(x) = \frac{dF(x)}{dx} \tag{1.19}$$

3. Moments

The nth moment about a specific real number C of a signal x, $E[(x - C)^n]$, is defined from the PDF as

$$E\left[(x-C)^n\right] = \int_{-\infty}^{\infty} (x-C)^n p(x)dx \tag{1.20}$$

where E is the mean or mathematical expectation. It is typical to calculate the moments about means, and luckily most of acoustic and audio signals have zero means. Therefore, it is most common to calculate moments about zero.

$$E\left[(x)^n\right] = \int_{-\infty}^{\infty} x^n p(x)dx \tag{1.21}$$

Obviously, for signals that span within a finite range of amplitudes, say $a < x < b$, the above becomes

$$E\left[(x)^n\right] = \int_{a}^{b} x^n p(x)dx \tag{1.22}$$

The first four moments are commonly used and known as mean, mean square (or variance when standardised), skewness, and Kurtosis. High-order moments, especially Kurtosis, is found to be a useful features for some audio signal classification applications. This will be discussed in later chapters.

So the mean, often denoted as μ, of a signal x is given by

$$\mu = E[x] = \int_{-\infty}^{\infty} xp(x)dx \tag{1.23}$$

And the mean square value is

$$E\left[x^2\right] = \int_{-\infty}^{\infty} x^2 p(x)dx \tag{1.24}$$

In statistics, standard deviation σ and variance σ^2 are often used to describe the distribution of random variables.

$$\sigma^2 = E\left[(x - E[x])^2\right] \tag{1.25}$$

For signals with zero means, the equation simplifies to give

$$\sigma^2 = E\left[x^2\right] - (E[x])^2 \qquad (1.26)$$

To put these in a nutshell, the standard deviation is the square root of the variance; the variance of the signal is the mean of its squares minus the square of its mean.

Some readers may wonder: Since these moments are all about the averages of the nth power of the signals, why bother integrating and using the PDF? Straightforward calculation of the nth power and arithmetic mean of discreet number series is possible and easier, but this is not applicable to continuous signals. Readers will see later that after digitisation of the continuous signals, many signal statistics become very simple.

1.4.2 Lag Statistical Analysis and Correlation Functions

The PDF and moments give insight into the statistical distribution of amplitudes of a signal. But two random processes with identical means and variances may well be very different. In addition to these, which are the first two moments that give us some ideas about how amplitudes change, we are also interested in how the past, current, and future states of a random signal are related statistically. That is to say, we are interested in the similarity of the signal and its time-shifted version. Lag statistics are used for this purpose. In essence, lag statistics look at the statistical feature of a signal and the time-shifted version of the signal itself. Lag statistics are used to identify dominating periodic or cyclostationary content in signals. The most commonly used lag statistic parameters are auto-correlation and cross correlation. The terminologies used are somewhat confusing; we go for an orthodox approach in defining them.

Auto-covariance function (ACVF), denoted as S_{xx}, is defined as the averaged value of the product of a signal and a time-shifted version of itself. This gives a measure of how the signal is similar to itself in the past. For an infinite duration, the ACVF is defined as

$$S_{xx}(\tau) = \lim_{T \to \infty} \frac{1}{T} \int_{-\frac{T}{2}}^{\frac{T}{z}} x(t)x(t+\tau)dt \qquad (1.27)$$

It is a measure of how $x(t)$ is correlated to itself after a delay of τ. T is the time window and the average is extended to infinity.

It is often useful to understand how two signals are related to each other as a function of time difference. The lag statistic of a signal can be extended to two signals $x(t)$ and $y(t)$ and, thus, we can define cross covariance function (CCVF) as

$$S_{xy}(\tau) = \lim_{T \to \infty} \frac{1}{T} \int_{-\frac{T}{2}}^{\frac{T}{z}} x(t)y(t+\tau)dt \qquad (1.28)$$

If the signal is time limited, say from $T1$ to $T2$, the equation above should be adjusted accordingly to become

$$s_{xx}(\tau) = \int_{T1}^{T2} x(t)x(t+\tau)dt \qquad (1.29)$$

and

$$s_{xy}(\tau) = \int_{T1}^{T2} x(t)y(t+\tau)dt \qquad (1.30)$$

The auto-correlation and cross correlation functions are probably the most commonly quoted. Sometimes they are used interchangeably or "confused" with the auto and cross covariances as defined above, i.e. some texts use the same definitions for auto-correlation and cross correlation. Nonetheless, the rigorous definitions for auto-correlation R_{xx} and cross correlation R_{xy} are the normalised S_{xx} and S_{xy} as

$$R_{xx}(\tau) = \frac{S_{xx}(\tau)}{S_{xx}(0)} \qquad (1.31)$$

and

$$R_{xy}(\tau) = \frac{S_{xy}(\tau)}{S_{xy}(0)} \qquad (1.32)$$

It is also useful to note from the definition of S_{xx} and the definition of $E[x^2]$ that

$$S_{xx}(0) = E\left[x^2\right] \qquad (1.33)$$

There are some other useful properties of these correlation functions:

- If the signal $x(t)$ has a periodic component, then its ACVF $S_{xx}(\tau)$ contains a periodic component of equal period.
- If either or both of the signals $x(t)$ and $y(t)$ are zero mean processes, then the CCVF will be zero if $x(t)$ and $y(t)$ are statistically independent.
- A non-zero value of the CCVF implies a linear relationship between $x(t)$ and $y(t)$.

Statistical features can also be acquired in the frequency domains. Some of the common frequency domain signal statistics will be summarised later in this chapter.

1.4.3 GAUSSIAN DISTRIBUTION AND CENTRAL LIMIT THEOREM

Gaussian distribution is of particular importance in probability theory and statistical signal processing; this is very much stated by the central limit theorem. The Gaussian distribution is often used in both natural and social sciences to model random variables with different underlying (unknown) distributions.

Gaussian distribution is the most commonly seen PDF for many physical processes. It is also called normal distribution. Its PDF is a bell-shaped curve shown in Figure 1.8. Recall Equation 1.17: the sum or integral over all values of x gives a probability of 1. The Gaussian distribution has a probability of 0.683 of being within one standard deviation of the mean. The probability density function p of the Gaussian process is given by

$$p(x \mid \mu, \sigma) = \frac{1}{\sigma\sqrt{2\pi}} e^{-\frac{(x-\mu)^2}{2\sigma^2}} \qquad (1.34)$$

The left hand side of the above function indicates $p(x)$ is determined completely by mean μ and standard deviation σ. It is interesting to note that for any Gaussian random signal, the first two moments, i.e. mean and variance, will be sufficient to completely characterise its statistical features. In fact, many random processes are assumed or modelled as Gaussian processes; in practice, only the first two moments need to be checked. Strictly speaking, this is only correct when the random process is Gaussian. However, the importance of Gaussian distribution, backed by the central limit theorem below, indicates that in handling most independent random variables when sample numbers are sufficiently large, one should obtain the average close to the Gaussian distribution.

The central limit theorem: Generally, the distribution of the sum (or arithmetic average) of a sufficiently large number of independent variables, with well defined (i.e. finite) means and variances, is Gaussian or approximately Gaussian, regardless of the underlying distribution.

1.5   SIGNALS IN TRANSFORMED FREQUENCY DOMAINS

Acoustic and audio signals are intrinsically one-dimensional time domain signals. Time dominion representation is obviously a straightforward format. Nonetheless, signals and system functions are often transformed in other domains mainly for two reasons: ease of manipulation and transparency of characteristics. Fourier transform and Laplace transform are by far the most prevalent ones in various applications. When we quote in a specific domain, we mean that both signals and system functions are transformed into that domain. That is to say, the same transform is applied to both the signals and the system functions.

1.5.1   FOURIER AND LAPLACE TRANSFORMS

Fourier transform by far is the most commonly used transform in signal and system analysis. The Fourier series expansion is used for harmonic analysis for periodical signals; the Fourier transform (Fourier Integral Transform), as a generalisation to the Fourier series, can be applied to non-periodical signals and the transform is invertible. This means that the Fourier transform is not a partial profile of the original signal; rather, it contains complete information of the original signal but represents it in a different format. By definition, the Fourier transform *FT* of a time domain function $x(t)$, denoted as $X(\omega)$, is defined by

$$X(\omega) = FT[x(t)] = \int_{-\infty}^{\infty} x(t)e^{-j\omega t}d(t) \qquad (1.35)$$

Fourier transform is invertible via the inverse Fourier transform *IFT*, given by

$$x(t) = IFT[X(\omega)] = \frac{1}{2\pi}\int_{-\infty}^{\infty} X(\omega)e^{j\omega t}d\omega \qquad (1.36)$$

It is straightforward to understand but important to note that while signals can be represented in the time domain or the frequency domain, the total signal energy should remain the same. This is known as Parseval's theorem and can be written as

$$\int_{-\infty}^{\infty} |x(t)|^2 \, dt = \frac{1}{2\pi}\int_{-\infty}^{\infty} |X(\omega)|^2 \, d\omega \qquad (1.37)$$

Laplace transform *LT* is closely related to Fourier transform.

$$X(s) = LT[x(t)] = \int_{0}^{\infty} x(t)e^{-st}dt \qquad (1.38)$$

and its inverse

$$x(t) = ILT[X(s)] = \frac{1}{2\pi}\int_{\infty}^{\infty} X(s)e^{st}ds \qquad (1.39)$$

where s as a complex number. If we restrict s to be imaginary, i.e. $s = jw$, and also choose $t \geq 0$, then the Laplace transform becomes the Fourier transform.

1.5.2 Signal Statistics in the Frequency Domain

Signal statistics can not only perform in the time domain but can also be applied in the frequency domain. One of the important purposes of performing the Fourier transform is to make the frequency (harmonic) components of the signal transparent. Power spectral density is used to check the signal power at different frequencies. This is achieved by applying lag statistics in the frequency domain.

The power spectral density (PSD) function of a wide sense stationary is defined as the Fourier transform of the auto co-variance function, i.e.

$$S_{xx} = PSD(\omega) = \int_{-\infty}^{\infty} s_{xx}(\tau)e^{-j\omega\tau}d\omega \qquad (1.40)$$

This means that PSD and ACVF are Fourier transform pairs! When checking the frequency contents of a random signal, we often use a loosely defined term "spectrum" of a signal. Evidently, the PSD is the result of lag statistics on the signals represented in the Fourier transform domain.

According to the Wiener-Khinchine theorem, an equivalent definition can be given to PSD:

$$PSD = S_{xx}(\omega) = \lim_{T \to \infty} \frac{1}{T} \left| \int_{-\frac{T}{2}}^{\frac{T}{2}} x(t)e^{-j\omega t} dt \right|^2 \tag{1.41}$$

In most cases, this means the power spectral density function of that signal. As we shall discuss later, the spectrum (PSD) of a random signal needs to be estimated in most times.

1.5.3 INPUT-OUTPUT RELATIONSHIPS OF LTI SYSTEMS

Recall the definition of a system in Figure 1.1 by reviewing Figure 1.9, which is a repeat of Figure 1.1. Also take into account that functions $x(t)$, $y(t)$, and $f(x)$ can all be represented in their native and transformed domains. We will look at how system input-output and system functions are linked in these domains.

The input to the system is $x(t)$ and the output after manipulation becomes $y(t)$; the system may be viewed as a function $f(x)$ that maps $x(t)$ onto $y(t)$. In the case that the system received only one input and yields one output, such a system is said to be a single-input-single-output (SISO) system. The mapping relation from input to output can take different forms depending upon how the system and signals are described. In the time domain, an LTI system is completely described or determined by its impulse response $h(t)$, which is the response one obtains from applying the Dirac's impulse signal as its input. This is shown in Figure 1.10 (top).

In the time domain, the system output $y(t)$ can be determined by the convolution of the input and the impulse response of the system.

$$y(t) = h(t) * x(t) = \int x(t - \tau)d(\tau)d\tau \tag{1.42}$$

Analytically, calculation of convolution on arbitrary complicated input signals is difficult. This is one of the many reasons why the frequency domains representations are used.

In the Laplace or complex frequency domain, a system can be described as Figure 1.10 (middle), in which $H(s)$ is the Laplace transfer function, which is the

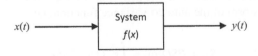

FIGURE 1.9   System definition (repeated Figure 1.1)

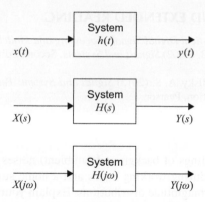

FIGURE 1.10 Signal and system in the time and transformed domains

Laplace transform of the system impulse response $h(t)$. A straightforward multiplication operation will find system output from its input and transfer function $H(s)$:

$$H(s) = LT[h(t)] = \int_0^\infty h(t)e^{-st}dt \qquad (1.43)$$

$$Y(s) = H(s) \cdot X(s) \qquad (1.44)$$

When the initial condition is zero, which for most acoustic and audio systems is the case, Laplace can be replaced by Fourier transform $s = j\omega$

$$Y(j\omega) = H(j\omega) \cdot X(j\omega) \qquad (1.45)$$

This is shown in Figure 1.10.

SUMMARY

Acoustic and audio signals are typically stochastic. Acoustic and audio signal systems are generally and preferably linear, unless psychoacoustic models are used to compress the signals. Most of the audio systems are linear and time invariant (LTI) ones. Even though minor weak nonlinearity and a small amount of time variance are commonplace in most real-world systems, in most cases these are negligible, and for the ease of analysis and design, LTI systems are often assumed. This chapter presented statistical signal modelling and description in both continuous time and frequency domains. The relationships between system input and output are established in the two domains. Some basic properties and characteristics of systems are also outlined. The content presented is self-contained. Most readers with a general science and engineering background should be able to follow without the need for texts. For those who are interested in rigorous mathematical proofs of the outlined properties of signal statistics and system modelling, a reading list is provided below as extended reading.

BIBLIOGRAPHY AND EXTENDED READING

Hayes, M. H. (2008) *Statistical Digital Signal Processing and Modelling*, Wiley.
Haykin, S. and Van Veen, B. (2002) *Signals and Systems*, Second Edition, John Wiley & Sons. Chapters 2, 3, and 6.
Oppenheim, A. V. and Willsky, A. S. (2013) *Signal and System: Pearson New International Edition*, Second Edition, Pearson.

EXPLORATION

1. Take some recordings of background (ambient) noises in a relatively big, open-plan office during working hours and a home study at a quiet time, and analyse their magnitude distributions. Explain your findings.

2. MATLAB® is a scientific computing platform and is commonly used for signal processing for algorithm development and fast prototyping. Some of the examples in this book are given in MATLAB codes. To start work in MATLAB, "MATLAB Primer" by MathWorks® is available online (https:// www.mathworks.com/help/pdf_doc/matlab/getstart.pdf) and is an invaluable learning resource. The best way of learning MATLAB is by using it. Read along while trying some of the example codes.

2 Sampling Quantization and Discrete Fourier

With the increasing applications of digital circuits and computers, it has been shown that there are advantages and cost savings in handing signals in their digital formats. As discussed in Chapter 1, signals are representations of physical quantities, and, thus, most signals are naturally continuous in both magnitude and time. Acoustic signals show no exception. To enable digital processing, signals first need to be converted to a digital format for processing, and if necessary, processed signals are reconstructed by converting their digital representations back to an analogue form. This is called analogue to digital conversion (ADC) and digital to analogue conversion (DAC). In this chapter, the principles of ADCs and DACs will be discussed first, with an emphasis on sampling and quantization, which are the two major steps to digitize signals. The essential relations and aspects of signals and systems as outlined in Chapter 1 will then be re-examined in the discrete time and completely digital domains. The discrete Fourier transform (DFT) will be outlined as the major frequency domain representation of digital signals.

2.1 SAMPLING

A physical quantity, e.g. sound pressure, is observed or measured, and represented to its analogue format, typically as a voltage signal. A general voltage signal, $v(t)$, can take any value at any point in time, i.e. it is continuous in both amplitude and time. In order to exactly record using data even a finite length segment of $v(t)$, an infinite volume of data must be stored for each of the amplitude and time domains, unless the signal is a simple deterministic one, in which case it can be represented by an equation. Fortunately, such enormous storage capacity is never practically required.

Example

Complicated moving visual images carry an infinite amount of detail. However, most of this detail is redundant, such that the information content can be maintained if the data is stored at low and finite resolution. This redundancy is exploited in video systems by subdividing the visual field (in the two spatial dimensions) into a number of discrete cells, called pixels. The colour spectrum and brightness of the light in each pixel is stored to finite resolution for digital transmission. Further, the frequency response of the human eye can be exploited to reduce the rate at which each pixel must be updated to give the image apparently smooth and continuous motion. The eye cannot perceive changes in intensity at frequencies much above 20 Hz due to the phenomenon of "persistence of vision." The pixels need not, therefore, be updated more often than once every 1/25th of a second.

This explains the digitisation process in a heuristic way. This process (in audio and video) involves two major and independent steps, namely quantization and discretization.

Quantization is the process that approximates a continuous amplitude domain variable as a value from a set of allowed quantum levels.

Example: In the video image, the original visual field is quantized spatially into pixels, and the colour balance and brightness information is quantized into a set of quantum levels.

Discretization is the process by which signals that are continuous in time are represented by a sequence of numbers describing the amplitude of the signal at discrete instants of time.

Example: In a moving video image, each pixel's brightness and colour balance needs to be updated once every $1/25^{th}$ second, rather than varying continuously with respect to time.

In digital signal processing (DSP) algorithm development and many parts of this book, we do not concern ourselves with amplitude quantization for two major reasons: (1) Many computational platforms, such as MATLAB®, handle sampled data as a very high precision floating point format so that the discrete representations are very close to their continuous format. (2) On occasions, when amplitude discretization becomes a concern, e.g. ADCs and DACs or fix-point calculation in certain DSP chips, the implications of amplitude discretization can often be modelled, and considered as quantization noise and finite word-length effects.

2.1.1 TIME DISCRETIZATION

The first stage of the "sampling" process is discretization, in which a function that varies continuously with respect to its independent variable is converted into a discrete sequence. Given a continuous time domain signal, perhaps a voltage $v(t)$, discretization represents the signal by a sequence of numbers, with each number recording the value of the function at a particular point in time. The time points chosen are usually regularly spaced—the signal is usually sampled at a constant "sampling frequency."

The sampling frequency is $f_s = 1/\Delta t$, where Δt is the sampling period or sampling interval T. Then the discrete time sequence representing $v(t)$ is $v_d(t)$. The discrete signal, v_k, is formed when the input signal $v(t)$ is multiplied in time by the sampling waveform $s(t)$. The relationship between $v(t)$, $s(t)$, and $v_d(t)$ is illustrated in Figure 2.1.

The waveform v_d is seen in the figure to have zero value for the vast majority of time. It can be stored efficiently by recording only the non-zero values, which occur at times $k\Delta t$ where k is an integer. These non-zero values, indexed by the integer k, form the terms of the sequence v_k, which is the time-discretized representation of $v(t)$.

As a consequence of the *multiplication* of the input signal $v(t)$ with the sampling waveform $s(t)$, the process of time discretization is *non-linear*. The sampling waveform used is a series of time-delayed impulses with unit magnitude (i.e. unit impulses); the time delay between each of these impulses is the sampling period.

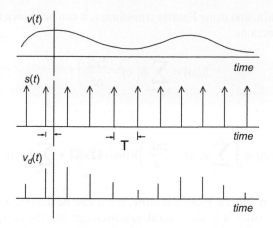

FIGURE 2.1   Sampling with a train of impulse at interval $T = \Delta t$

Another way of thinking of this is that the error between the sampled waveform v_d and original waveform v is signal dependent, and signal dependent errors are, therefore, distortion.

This non-linearity generates frequency components in the quantized representation of the data v_k that are not present in the original signal $v(t)$. If care is not taken with the specification of the discretization process, then the data can be irretrievably corrupted; this corruption is due to the phenomenon of *aliasing*, which we shall discuss later.

The accuracy of the discretizing process is dictated by the ratio between the sample frequency and the frequency of the input signal. If the sample frequency is greater than twice the input signal's highest component, then no corruption occurs—this is Shannon's theorem.

2.1.2 ALIASING

The multiplication of the continuous time signal with the sampling waveform is related to the convolution of the input signal's frequency domain representation with the frequency domain representation of the sampling waveform. In order to fully understand the time-discretising process, we must consider the effects of this convolution.

The discretisation operation in time is based upon multiplication:

$$v_d(t) = v(t).s(t) = v(t).\sum_{k=-\infty}^{\infty} \delta(t - k\Delta t) \tag{2.1}$$

where the sampling waveform is a train of delta functions $\delta(t)$.

This can be represented in the frequency domain by the convolution of the Fourier transform of the voltage signal, $V(\omega)$, and the Fourier transform of the sampling waveform, $S(\omega)$:

$$V_d(\omega) = V(\omega) * S(\omega) \tag{2.2}$$

$s(t)$ was a pulse train, and using Fourier transforms, it can be shown that $S(\omega)$ is also a train of delta functions:

$$S(\omega) = \sum_{k=-\infty}^{\infty} \delta\left(\omega - \frac{2\pi k}{\Delta t}\right) \tag{2.3}$$

Expressing the convolution in integral form, it can be written as:

$$V_d(\omega) = \int_{-\infty}^{\infty} \sum_{k=-\infty}^{\infty} \delta\left(\Omega - \frac{2\pi k}{\Delta t}\right) V(\omega - \Omega) d\Omega = \sum_{k=-\infty}^{\infty} V_{d,k}(\omega) \tag{2.4}$$

This convolution integral looks daunting, but it can be understood if it is considered one term at a time. It is also useful to remember that the sampling frequency $f_s = 1/\Delta t$. For any one of the integer values of k, a component of the Fourier transform of $V_d(T)$ results:

$$V_{d,k}(\omega) = \int_{-\infty}^{\infty} \delta(\Omega - 2\pi k f_s) V(\omega - \Omega) d\Omega \tag{2.5}$$

which can be simplified to:

$$V_{d,k}(\omega) = V(\omega + 2\pi k f_s) \tag{2.6}$$

using the following shifting property for delta functions:

$$\int f(x) \delta(x - a) dx = f(a) \tag{2.7}$$

Each of these $V_{d,k}(T)$'s in Equation 2.6 represents the original voltage signal, shifted in frequency up to the kth harmonic of the fundamental sampling frequency. Figure 2.2 shows, for example, the magnitudes of a typical $V(T)$ and the associated $V_d(T)_0$ and $V_d(T)_1$.

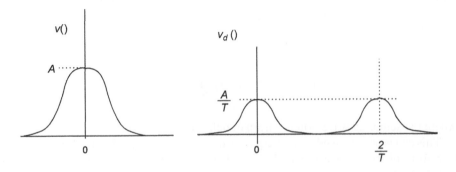

FIGURE 2.2 Spectra of the original continuous signal (left) and the first two repeats of the discrete time sampled version (right)

The total Fourier transform, $V_d(T)$, of the discrete time series $V_d(t)$ is formed of the sum of all the individual elements above:

$$V_d(\omega) = \sum_k V_{d,k}(\omega) \qquad (2.8)$$

and the Fourier Transform of a discrete time series is *periodic in frequency*, with a period of f_s. Because of the periodicity of $V_d(T)$, it is not necessary to record all of the function; if one period is recorded, the whole function is known. It is, therefore, only necessary to record data over the baseband interval:

This reduction in the volume of information needed to specify v is the exact frequency domain analogue of the data reduction inherent in time domain discretization, in which the continuous signal's amplitude needs to be recorded only once every Δt seconds.

The redundancy of the periodic repeats in the frequency domain representation of v_d can also be exploited to define a new frequency transform appropriate to discrete-time signals and systems, the z-transform.

Generally, $V_d(T)$ within the baseband interval $-f_s/2$ to $f_s/2$, is formed from additive contributions from several $V_{d,k}(T)$'s both within and outside this frequency range. This phenomenon is called *aliasing* as frequency components of $V(T)$ can *masquerade as components of $V_d(T)$ at different (lower) frequencies*. Once this aliasing has occurred, the data is corrupted and the original information can never be restored.

A common example of aliasing is when the spokes of wagon wheels in western movies often appear to rotate slowly backwards when the wagon is travelling forward at speed. The continuous moving image is time-discretized into film frames (f_s of order $1/25^{\text{th}}$ second) and the slow backwards motion is caused by a high (negative) frequency component of one of the sidebands of the first harmonic of the frame frequency falling within the visible baseband.

2.2   FOURIER

The function $s_n(x)$:

$$s_n(x) = \frac{a_0}{2} + \sum_{k=1}^{n} a_k \cos[2\pi kx / L] + b_k \sin[2\pi kx / L] \qquad (2.9)$$

is periodic with period L, since it is the sum of harmonically related components, the fundamental ($k = 1$), which has a period equal to L. The function $s_n(x)$ in Equation 2.9 is called a trigonometric polynomial, order n, period L. Fortunately for us, many periodic signals found in audio and acoustics may be usefully approximated as high order trigonometric polynomials (or accurately described by infinite order series). In describing a signal in terms of an equivalent trigonometric polynomial, we are stating that combination of harmonic components that represents the signal, i.e. we are performing *frequency analysis*.

Although the above cosine and sine representation seems more intuitive, a complex representation is far more useful: The complex exponential form of the Fourier series:

$$f(x) = \sum_{n=-\infty}^{\infty} c_n e^{2n\pi jx/L}$$

(2.10)

where the (complex) coefficients c_n are calculated by:

$$c_n = \frac{1}{L} \int_{-L/2}^{L/2} f(x) e^{-2\pi njx/L}$$

(2.11)

If we plot the value of the magnitude of c_n against frequency index or frequency, we get a graphical description of the frequency content of the signal under study—a *spectrum*. Since the magnitude of c_n is only non-zero for certain values of frequency (corresponding to the integer values of n), the picture we produce is a *line spectrum*.

Two examples of line spectra ($|c_n|$) of important periodic signals are shown below: the spectrum in Figure 2.3a is for a sinusoidal wave; the spectrum in Figure 2.3b is for a pulse train of delta functions. The different signals have clearly different frequency contents; they can be synthesised by combining harmonics in different ratios. As an additional example, Figure 2.3c shows a more complicated line spectrum, which is the one of a square wave.

The contribution of each harmonic to the Fourier series corresponds to the magnitude of the complex coefficient $|c_n|$. c_0 is *twice the mean value of f(x):*, this might also be referred to as the zero-frequency or dc component of the signal.

In order to specify the ratios of the complex exponential components of each signal, we would need more than the magnitude expressed in the line spectra above. This is given by the *phase* of each complex exponential component.

2.3   FOURIER SERIES OF PERIODIC, DISCRETE-TIME SIGNALS

Now consider a sequence of numbers, x_k, that is periodic:

$$x_k = x(k\Delta t) = x(k\Delta t + P)$$

(2.12)

where Δt is the sampling period; k is the (integer) time index, and P is the period ($P = n\Delta t$, n is an integer)

Perhaps x_k was generated by taking a continuous time signal, $x(t)$, and time discretising it. The time discretisation is achieved by multiplying the continuous signal by a "sampling" waveform, which is non-zero only at those instants of time when $x(k\Delta t)$ is to have value. Such a sampling waveform is a delta function train:

$$x_d(t) = \sum_{k=-\infty}^{\infty} \delta(t - k\Delta t) \cdot x(t)$$

(2.13)

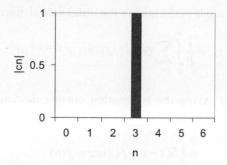

FIGURE 2.3a Spectrum of sinusoidal wave

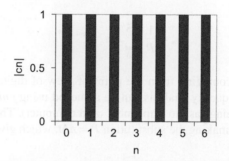

FIGURE 2.3b Spectrum of pulse train

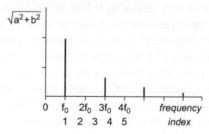

FIGURE 2.3c Line spectrum of a square wave

If we wish to *synthesise* the discrete time signal, we should anticipate the use of an infinite number of harmonics to adequately represent the "spikes" introduced by the sampling waveform, so we introduce Equation 2.14:

$$x_d(t) = \sum_{m=-\infty}^{\infty} c_m e^{j2\pi mt/P} \tag{2.14}$$

Notice that this expression actually synthesises a continuous time signal—it does not limit k to be an integer. If we only wish to synthesise the values of the sequence x_k (the values of $x(k\Delta t)$ when k has an integer value), then perhaps the infinite number of harmonics in the expression above will not be required—we shall see later on!

Fourier analysis of $x(t)$ is performed in the usual way (defined by Equation 2.14):

$$c_m = \frac{1}{P}\int\limits_0^P \sum_{k=0}^{n-1} \delta(t - k\Delta t) \cdot x(t)e^{-j2\pi mt/P}\, dt \tag{2.15}$$

which we simplify by taking the summation outside the integral sign and then exploiting the "sifting property" of delta functions:

$$\int \delta(x - a) \cdot f(x)\,dx = f(a) \tag{2.16}$$

This gives:

$$c_m = \frac{1}{P}\sum_{k=0}^{n-1} x_k e^{-j2\pi mk\Delta t/P} \tag{2.17}$$

which looks very encouraging from the point of view of digital signal processing as it suggests that frequency analysis can be achieved using *finite sums* rather than *integrations* (sums being easy to implement on a computer). The expression can be further simplified by making the substitution $P = n\Delta t$, which gives:

$$c_m = \frac{1}{P}\sum_{k=0}^{n-1} x_k e^{-j2\pi mk/n} \tag{2.18}$$

The expression above is very useful as it has very little dependence upon time; P is the only element of the expression that has time units. This allows Equation 2.18 to be used to study the "frequency" content of number sequences that are derived from any source; the "frequency" analysis then focusses on how the sequence varies from sample to sample, rather than how a signal varies in time.

We have seen, from consideration of Fourier synthesis of $x_d(t)$, that an infinite number of c_m's are generated in the frequency analysis of discrete-time periodic signals. If, however, we look at the Fourier analysis expression for c_m in Equation 2.18, we see that the c_m's are a periodic sequence in frequency having period n. It is then not necessary to store all of the c_m's in order to understand the frequency content of the discrete-time signal—we need only store the first n. This might start to sound familiar. In section 2.2, we described how a time discretised digital signal was periodic in frequency

These n harmonics, weighted by the appropriate complex coefficients c_m, describe the n values of the sequence x_k, in the sense that these n values could be synthesised by Fourier synthesis using only a finite number of harmonics.

Equation 2.18 represents a discrete Fourier transform (DFT). Although the DFT opens an enormous range of new computing opportunities, allowing frequency domain analysis to be applied to all kinds of data, the real revolution in digital signal processing did not take place as a result of the DFT, but due to a more efficient algorithm called the fast Fourier transform (FFT). The FFT breaks down an $N = 2^n$ length DFT into a series of shorter DFTs (in the end N length 1 transforms), and, therefore, operates considerably faster than a straight application of Equation 2.14.

It is important to emphasize that the FFT is designed only to generate the same set of frequency domain coefficients c_m from an input vector x_k as the DFT; it is simply an algorithm, which incurs lower computational cost. Where the DFT has a cost of N^2, the cost of the FFT is $N\log_2 N$. It is the FFT algorithm that performs the DFT calculations at the heart of all modern digital instrumentation used in audio and electroacoustics. The frequency bandwidth that is covered by the discrete transforms generated by the FFT is dictated solely by the sample rate if aliasing is avoided by following Shannon's sampling theorem:

$$\text{FFT/DFT bandwidth is } 0 \text{ Hz} - 0.5\Delta t^{-1}\text{Hz} \dots \text{ or}$$

$$\text{FFT/DFT bandwidth is } 0 \text{ Hz} - 0.5 f_s \text{ Hz.}$$

The frequency resolution with which this bandwidth is covered is controlled by the order of the transform, n, which is ultimately limited by the record length:

$$\text{FFT/DFT resolution is } (n\Delta t)^{-1}\text{Hz.}$$

POSITIVE AND NEGATIVE FREQUENCIES

If the first harmonic ($m = 1$) is understood to be at the fundamental frequency of the original signal, then the range of n values of c_m correspond to a frequency range from:

$$0 \text{ to } \left(1 - n^{-1}\right) \cdot \Delta t^{-1}$$

which (for large n) means the frequency range goes up to $\sim 1/\Delta t$.

At first sight, this may appear to contradict Shannon's sampling theorem (as there should be no frequency content of the original signal above $0.5/\Delta t$). If, however, we remember the periodicity of the c_m's, then those values of c_m lying in the interval $0.5/\Delta t$ to $1/\Delta t$ are equal to c_m's with negative values of m, corresponding to negative frequencies in the interval 0 to $-0.5/\Delta t$. This means that our n coefficients refer to frequencies in the range

$$-0.5/\Delta t \text{ Hz to } 0.5/\Delta t \text{ Hz}$$

which suggests no problem with Shannon's sampling theorem, as we shall see later. When a fast Fourier algorithm is used, results for negative frequencies are produced. Acoustically, these can, with caution, be seen as an anomaly, but care must be taken as these contain half the power of the acoustic signal, as we shall see later.

2.4   PRACTICAL FFTs

Load up MATLAB. You will need the script file *fourierdemo.m*. Run this file. The MATLAB script *fourierdemo.m* creates a sinusoidal wave, takes the FFT, and displays the power spectrum.

2.4.1   POSITIVE AND NEGATIVE FREQUENCIES

Look first at MATLAB Figure 1 (plot generated from the demo code). We get two dominant spikes in the frequency response: one in bin number 38, the other in bin 988. This relates to the point in the box "Positive and Negative Frequencies" above. The first spike is the 400 Hz input signal. The second is actually a −400 Hz (minus 400) peak—the negative frequency.

Look at MATLAB Figure 2 (plot generated from the demo code), where the positive and negative frequencies are shown in magnitude and phase. Note the symmetry. The Fourier transform (Y) has the following property:

$$|Y(f)| = |Y(-f)|$$
$$Y(f) = Y^*(-f)$$

(2.19)

where * denotes complex conjugate. So while 1000 and −1000 Hz have the same magnitude, they are 180 degrees out of phase. In the MATLAB command window, type "Y(94)" and "Y(932)" to look at the contents of these cells, which relate to 1000 and −1000 Hz respectively.

In acoustics, negative frequencies are not that meaningful. But we have to be careful because half the energy of the spectrum is contained there. Look at the output generated by lines 53-55. If we only use the positive frequencies, we lose 3dB from the spectra. When overall power is not important, we ignore the negative frequencies in acoustics. Table 2.1 gives a summary of how data are arranged in the time and frequency domains.

TABLE 2.1
How data are arranged after an FFT from time to frequency

Bin Number	Frequency	
1	0 (dc)	Positive Frequencies
2	$1/N\Delta t$	
3	$2/N\Delta t$	
...		
$N/2$	$(N/2 - 1)/N\Delta t$	
$N/2 + 1$	$1/2\Delta t$ i.e. Nyquist and $-1/2\Delta t$	Positive and Negative
$N/2 + 2$	$-(N/2 - 1)/N\Delta t$	Negative
...		
$N - 1$	$-2/N\Delta t$	
N	$-1/N\Delta t$	

Note that when you do an inverse FFT from frequency to time, *you must have both positive and negative frequency information*. Consider the inverse DFT:

$$x_k = \sum_{m=0}^{n-1} c_m e^{j2\pi mk/n} \tag{2.20}$$

In general, this transform forms complex components, which is not useful because proper time signals must be purely real. Arranging the c_m coefficients to have even (with respect to frequency) real parts and odd imaginary parts forces the transform to produce only real values for x_k as the imaginary components cancel. MATLAB Figure 3 illustrates this. When positive and negative frequencies are included (top graphs), we get a purely real output (note that the y-scale on the imaginary part is very small). If we include only the positive frequencies (bottom), things go wrong and we get a significant imaginary time signal, which has no physical meaning. We shall return to this later in Chapter 4 in finite impulse response (FIR) filter design.

2.4.2 WINDOWING

Return to MATLAB Figure 2. Now we put in a pure tone sine wave and, as expected, one frequency dominates. But there is significant width to this peak, i.e. other frequencies are present. Why does this occur?

To define a frequency exactly needs an infinite number of periods; otherwise, there is uncertainty in the measurement. *In the Fourier transform method of power spectral density estimation, we have to use* truncated Fourier transforms—*we cannot measure a signal forever*. The consequence of using a truncated Fourier transform is that we generate a power spectral estimate *not* of the original signal, but of a time windowed version of the signal. The true Fourier transform and the truncated Fourier transform are different. To understand the error, we must first reintroduce an important theorem of Fourier transforms. Examples of a signal and a truncated version of that signal are shown in Figures 2.4a and 2.4b.

2.4.3 THE CONVOLUTION THEOREM

Given two time domain signals, $x(t)$ and $w(t)$, the Fourier transform of the product, $x(t).w(t)$, is given by the convolution of the individual Fourier transforms of x and w:

$$FFT\{x(t) \cdot w(t)\} = FFT\{x(t)\} \otimes FFT\{w(t)\} \tag{2.21}$$

where $\otimes$ denotes convolution.

Returning to our truncated Fourier transforms and time windowed data, we must now consider how the time windowing process is achieved. The truncated signal $x_T(t)$ is related to the original signal $x(t)$ by multiplication by a windowing function, $w(t)$, such that:

$$\begin{aligned} w(t) &= 1 & -T/2 \leq t \leq T/2 \\ &= 0 & elsewhere \end{aligned} \tag{2.22}$$

$$x_T(t) = x(t) \cdot w(t) \tag{2.23}$$

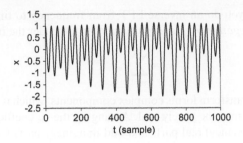

FIGURE 2.4a   Original signal

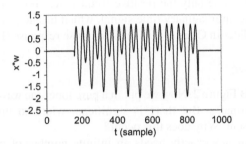

FIGURE 2.4b   Truncated signal

The Fourier transform of the truncated signal can now be derived using our understanding of the convolution theorem. The Fourier transform of a time windowed signal is the true Fourier transform of the signal convolved with the Fourier transform of the windowing signal, *W*.

The Fourier transform of the rectangular window is a familiar problem:

$$
\begin{aligned}
W(\omega) &= \int_{-\infty}^{\infty} w(t)e^{-j\omega t}\, dt \\[2mm]
&= \int_{-T/2}^{T/2} e^{-j\omega t}\, dt \\[2mm]
&= T\frac{\sin(\omega T/2)}{\omega T/2} \\[2mm]
&= T\, SINC(\omega T/2)
\end{aligned}
\tag{2.24}
$$

where *SINC*() function is sin(*a*)/*a*. (Sometimes, for example in MATLAB, *SINC* is defined as sin(π*x*)/(π*x*)). The Figure 2.5 shows the Fourier transform of a rectangular windowing signal of length *L*. The window is symmetrical about *t* = 0 for simplicity.

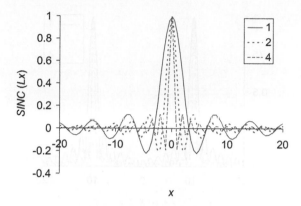

FIGURE 2.5   SINC functions

The $\sin(a)/a$ function has a major lobe that has width $4\pi/T$; it gets narrower as the window length gets wider. The height of the major lobe is proportional to the record length, so as the window length gets longer and longer, its Fourier transform tends towards a delta function (multiplied by 2π).

In the limiting case where the window becomes infinitely wide, the truncated Fourier transform X_T approaches the true Fourier transform X. When the window length is shorter, the window's Fourier transform is broader and the Fourier transform of the windowed signal X_T becomes less like the Fourier transform of the signal X.

We shall study how X and X_T differ by considering a cosinusoidal signal, the Fourier transform of which we know to be the sum of two delta functions. If a cosinusoid is windowed by our rectangular window signal, centred on $t = 0$, the Fourier transform of the truncated signal is (exploiting Equations 2.21 and 2.24):

$$X_T(\omega) = FFT\left(x_t\right)$$

$$= FFT(x) \otimes FFT(w)$$

$$= \left(\delta\left(\omega - \frac{2\pi}{T}\right) + \delta\left(\omega + \frac{2\pi}{T}\right)\right) \otimes T\, SINC(\omega T/2)$$

(2.25)

The delta functions simply cause a shift in the *SINC* function to +/− the frequency of the cosinusoid, as is illustrated in the figure below.

The Figure 2.6 shows a series of power spectral density estimates of a cosinusoid. The PSDs were estimated using the Fourier transform method, such that the estimates are proportional to the magnitude squared of the equation above. The cosinusoid's frequency is 10 rad/sec and the window lengths are 1, 2, and 4 seconds ($L = T$) (approximately 1.5, 3, and 6 cycles of the cosinusoid). When the window length is short, the power in the signal is *smeared* to other frequencies in the PSD estimate. As the record length is increased, the *spectral smearing* is reduced and the power is more accurately identified as lying at +/−10 rad/sec.

The consequence of time windowing is to smear the frequency content of signals during the estimation of frequency domain statistics.

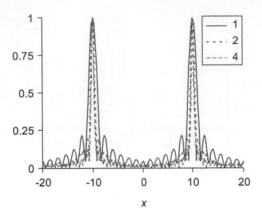

FIGURE 2.6   The PSD of cosine functions

The smearing of frequency domain statistics, imposed by the application of time domain windows, has the effect of introducing a finite *frequency resolution* to our analysis, as you saw in fourierdemo.m. We have noted that the major lobes of the $\sin(x)/x$ functions in the figure above have width $4\,\pi\,T^{-1}$ rad/sec (i.e. a frequency bandwidth of $2T^{-1}$ Hz). This means that PSD estimates produced by the Fourier transform method using time windowed data have been effectively produced by a bandpass filtering method where the bandpass filter has a magnitude frequency response described by a $\sin(x)/x$ function.

It is customary to define such an analysis bandwidth in terms of the −3dB bandwidth of the filters (in this case, the Fourier transform of the windowing signal). Using this approach, the −3dB points of the (squared) $\sin(x)/x$ filter responses is defined by $2^{1.5}/\pi T$.

Note: In the context of digital signal processing (DSP), the maximum frequency is dictated by Shannon's sampling theorem and the frequency resolution is controlled by the window length used in collecting the data.

2.4.4   AVOIDING SPECTRAL SMEARING—MORE WINDOWS

The spectral smearing in the examples discussed was caused by the shape of the Fourier transform of the rectangular windowing function. It may be worthwhile to consider a different windowing shape, which still limits the data to zero outside the interval $|t| > T/2$, and yet preserves some of the information of the original signal inside that interval with less frequency smearing. Remembering, from our recent experiences of rectangular windows, that the lowest amount of spectral smearing occurred when the window's Fourier transform looked most like a delta function, we should expect that good windows will have "narrow" major lobes and small side lobes. To identify candidates for new windowing functions, we should first recall the "inverse spreading" principle of Fourier transforms:

Features of time domain signals that change rapidly with time (ie "narrow" time domain features) will cause components of the Fourier transform that change slowly with frequency (i.e. are "broad" in frequency)

The narrow rapidly changing features of the rectangular window giving rise to its "wide" Fourier transform are the transitions at either end, so new candidate window functions should avoid such rapid jumps.

A potential candidate is the half cosine window:

$$w(t) = \cos\left(\frac{\pi t}{T}\right) \quad |t| < T/2 \tag{2.26}$$

This half cosine window has a broader central lobe but much smaller side lobes as compared to the rectangular window's Fourier transform.

A better window still is the "raised cosine" or "Hanning" window:

$$w(t) = \frac{1}{2} + \frac{1}{2}\cos\left(\frac{2\pi t}{T}\right) \quad |t| < T/2 \tag{2.27}$$

EXERCISE FOR MATLAB

Run the script *fourierdemo.m*. Compare the spectrum produced with and without a Hanning windowing applied. You should see a dramatic reduction in leakage, i.e. the power should be more concentrated on the 400 Hz frequency bin. Also, look at the time signal before and after windowing. Notice that the window smoothes the signal to zero at $t = 0$ and $t = T$, and, hence, prevents sudden truncation between the part where the signal exists—$|t| \leq T/2$—and the part where it does not—$|t| > T/2$. Also calculate the power in the windowed and non-windowed spectrum: Does windowing affect the total power measured?

People can get very worked up about the type of window to use: Hanning, Hamming, Parzen, Bartlett, Welch, etc. The important thing is to use a window of some type to reduce truncation (where appropriate). There may be very critical applications where different anti-truncation windows give significantly different results, but these situations are rare.

When to use a window:

1. If you have a continuous signal of which you are sampling a section, then an anti-truncation window such as Hamming should be applied.
2. If you are sampling a transient signal that starts and ends in the time frame $|t| \leq T/2$, then a rectangular window is automatically applied.
3. If you are sampling a transient signal that starts after $t > -T/2$ but ends after $t > T/2$, then half windows should be applied.

This is illustrated by the MATLAB script *windows.m*. A decaying sine wave is to be analysed (top trace in MATLAB Figure 1). Applying a Hanning window (middle) is obviously wrong because it removes the most important part of the signal. In this case, a half Hanning (bottom) is best because it prevents truncation errors at the right side of the signal, while not altering the left side. The windows are shown in MATLAB Figure 2.

This window has no step discontinuity at its edges and also has no step discontinuity in its gradient at the edges, and so has even lower side lobes than the half cosine window, at the expense of a slightly wider major lobe.

2.5 ESTIMATING STATISTICS USING FOURIER METHODS

You will already have seen in Chapter 1 that it is possible to estimate the power spectrum using the Fourier transform method. We can also use the same methods to estimate important statistics such as the auto-covariance function and cross covariance function.

We have seen that it is possible to make a decomposition of the statistics of a *signal* in amplitude, time, or frequency. In this module, we shall turn our attention to descriptions of the performance of *systems*. The systems will be assumed to be linear and to have a performance that does not change with time—linear time invariant (LTI) systems. We have seen that it is possible to represent any signal in terms of the amplitude and phase of the signal's constituent simple harmonic components (using Fourier techniques). An LTI system will respond to each of the Fourier components presented at its input *independently*; each Fourier component at the input will generate a component at the output *at the same frequency*. The system simply causes the Fourier components of the input to be scaled in amplitude and shifted in time before passing them to the output—the system has a *gain* and *phase* associated with each frequency.

We conventionally combine the gain and phase into a single complex valued function of frequency, called the frequency response, $H(\omega)$, in which

$$Gain = |H(\omega)|$$
$$Phase\ angle = \angle H(\omega)$$

(2.28)

At any frequency, the action of the system in producing output Y in response to input X is expressed by

$$Y(\omega) = X(\omega) \cdot H(\omega)$$

(2.29)

in which X and Y are the Fourier transforms of the input and output signals, respectively. Unfortunately, the Fourier transform is impossible to calculate for practical signals, so, although the input-output relationship above for an LTI system is perfectly correct, it has little practical value. Practically speaking, we are forced to make *best estimates* of transfer functions using estimated statistics of the input and output signals. This estimation (mainly applied in the frequency domain) is the subject of the next few sections.

2.5.1 CROSS POWER SPECTRAL DENSITY FUNCTION

Just as the power spectral density gives a frequency domain decomposition of the auto-covariance (through the Wiener-Khinchine theorem)

$$S_{xx}(\omega) = \int_{-\infty}^{\infty} s_{xx}(\tau) e^{-j\omega\tau}\, d\tau$$

(2.30)

it is possible to define a similar decomposition of the cross covariance

$$S_{xy}(\omega) = \int_{-\infty}^{\infty} s_{xy}(\tau)e^{-j\omega\tau}\,d\tau \tag{2.31}$$

where the function S_{xy} is the cross power spectral density (CPSD) function between the signals x and y. The CPSD is a frequency domain expression of the degree of *linear relationship* between x and y. Therefore, the CPSD has major application in the measurement (or, more properly, estimation) of linear transfer functions.

Example

Consider the two signals: $x = A\cos(\omega_0 t)$, $y = B\sin(\omega_0 t)$
The cross-covariance can be calculated in time from:

$$s_{xy}(\tau) = \frac{AB}{T}\int_0^T \cos\big(\omega_0 t\big)\sin\big(\omega_0(t-\tau)\big)d\tau \tag{2.32}$$

Remembering that $T = 2\pi/\omega$

$$s_{xy}(\tau) = \frac{AB}{2}\sin(\omega_0\tau) \tag{2.33}$$

Notice that the cross covariance has the same shape as the auto-covariance of a cosinusoid (which is a cosinusoid) shifted in lag (because the sinusoid is a cosinusoid shifted in time).
The CPSD is obtained by Fourier transforming the cross-covariance function:

$$S_{xy}(\omega) = \frac{AB}{2j}\Big[\pi\delta\big(\omega-\omega_0\big)+\pi\delta\big(\omega+\omega_0\big)\Big] \tag{2.34}$$

Note that the CPSD is, in general, a *complex* function of frequency.

2.5.2 ESTIMATING THE CPSD

Just as we had problems calculating power spectral density functions for anything but simple deterministic signals, we need some technique for *estimating* CPSDs. Estimating CPSDs uses the Fourier transform method, familiar from estimating auto power spectra, in which "raw estimates" formed by the product of truncated Fourier transforms are averaged:

$$S_{xy}(\omega) = \frac{1}{N}\sum_{i=1}^{N}\frac{X_i(\omega)Y_i^*(\omega)}{T} \tag{2.35}$$

where the truncated Fourier transforms are calculated over the same time intervals to preserve phase synchrony between the signals.

To simplify the notation, we shall introduce the idea of the *expected value* operator, $E[]$, which returns the averaged value of the operand inside the brackets:

$$E\left[X_T(\omega)Y_T^*(\omega)\right] = \frac{1}{N}\sum_{i=1}^{N}X_i(\omega)Y_i^*(\omega) \tag{2.36}$$

$$\Rightarrow S_{xy}(\omega) \approx E\left[\frac{X_T(\omega)Y_T^*(\omega)}{T}\right] \tag{2.37}$$

In order to begin to understand the application of the CPSD function, consider the CPSD function between the input x and output y of a linear system, H.

$$\Rightarrow S_{xy}(\omega) \approx E\left[\frac{X_T(\omega)Y_T^*(\omega)}{T}\right] = E\left[\frac{X_T(\omega)\left(X_T(\omega)H(\omega)\right)^*}{T}\right] \tag{2.38}$$

Since the transfer function H is a fixed property of the system, we may remove it as a fixed factor from the expected value operator, leaving:

$$S_{xy}(\omega) = H(\omega)S_{xx}(\omega) \tag{2.39}$$

This result explains the answer we found when we considered the CPSD between a cosinusoid and a sinusoid in the example above. In that case, the "magnitude" of the transfer function relating the two signals was B/A and the phase is associated with the 90° shift between sin and cos. The product of the auto-power spectrum of a sinusoid with this transfer function gives the results found for the CPSD.

It appears from the equation above that the CPSD function should have application in transfer function measurement. The CPSD is actually more powerful than is suggested by our example above, as demonstrated in the example below, but first a summary in Table 2.2:

TABLE 2.2
PSD Summary

Auto-spectra	$\Rightarrow S_{xx}(\omega) \approx E\left[\dfrac{X_T(\omega)X_T^*(\omega)}{T}\right]$	(2.40)
	where X_T is the FFT(x_T)	
Auto covariance	$S_{xx}(\tau) = \text{IFFT}\left(S_{xx}(\omega)\right)$	
Auto correlation function	$R_{xx}(\tau) = S_{xx}(\tau) / S_{xx}(0)$	
Cross spectra	$\Rightarrow S_{xy}(\omega) \approx E\left[\dfrac{X_T(\omega)Y_T^*(\omega)}{T}\right]$	(2.41)
	where X_T is the FFT(x_T) and Y_T is the FFT(y_T) and * denotes complex conjugate	
Cross covariance	$S_{xy}(\tau) = \text{IFFT}\left(S_{xy}(\omega)\right)$	
Cross correlation function	$R_{xy}(\tau) = S_{xy}(\tau) / S_{xy}(0)$	

2.6 TRANSFER FUNCTION MEASUREMENT IN NOISE

Consider the following Figure 2.7, which is typical of real-world observation of the linear system H. When an input x is applied, we cannot observe the response associated with that input; we are always faced with some amount of contaminating noise on the output, n.

If we calculate the CPSD function between input and observed response (which includes the inevitable noise), we get:

$$\Rightarrow S_{xy}(\omega) \approx E\left[\frac{X_T Y_T^*}{T}\right] = E\left[\frac{X_T \left(X_T H + N_T\right)^*}{T}\right] \qquad (2.42)$$

If we assume that the signals and the unwanted noise are independent of X, the CPSD simplifies as X and N produce zero correlation:

$$\Rightarrow S_{xy}(\omega) = E\left[\frac{X_T \left(X_T H\right)^* + X_T N_T^*}{T}\right] = E\left[\frac{X_T \left(X_T H\right)^*}{T}\right] = S_{xx}(\omega)H(\omega) \quad (2.43)$$

This important result shows that CPSD functions allow us to estimate linear transfer functions even in noise.

In summary:

- When two signals are uncorrelated (as in the example above when $H = 0$), their CPSD function has zero value.
- When two signals are perfectly correlated (as in the example above when $n = 0$), the CPSD's magnitude takes highest value.
- When two signals are only partially correlated (as in the example above when H and n are non-zero), the CPSD function has intermediate value.

2.6.1 THE ORDINARY COHERENCE FUNCTION

We have seen that when there is no correlation between two signals, the CPSD function between them has zero value. Conversely, when two signals x and y are perfectly correlated:

$$Y(\omega) = X(\omega) \cdot H(\omega) \qquad (2.44)$$

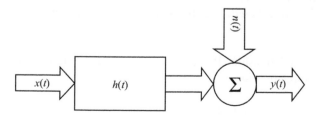

FIGURE 2.7 Observation of a linear system

so their CPSD takes value:

$$S_{xy}(\omega) = H(\omega)S_{xx}(\omega) \tag{2.45}$$

The magnitude squared CPSD function for such perfectly correlated signals is:

$$|S_{xy}(\omega)|^2 = |H(\omega)|^2 |S_{xx}(\omega)|^2$$
$$= S_{xx}(\omega)S_{yy}(\omega) \tag{2.46}$$

These two cases represent the limiting values that the magnitude squared CPSD function can assume, and these limits are expressed by Schwartz's inequality:

$$0 \le |S_{xy}(\omega)|^2 \le S_{xx}(\omega)S_{yy}(\omega) \tag{2.47}$$

We can quantify how close to perfectly correlated or perfectly uncorrelated two signals are, as a function of frequency, using a statistic derived from Schwartz's inequality. This statistic is called the ordinary coherence function (usually simply called coherence) and is defined as follows:

$$\gamma_{xy}^2(\omega) = \frac{|S_{xy}(\omega)|^2}{S_{xx}(\omega)S_{yy}(\omega)} \tag{2.48}$$

It is evident from Schwartz's inequality that the coherence function can range between zero and unity. The coherence function is a measure of the proportion of the *power* in *power* that is due to linear operations on the signal x. When estimating transfer functions, the coherence function is a useful check on the quality of the data used. In our example of estimating a transfer function in the presence of additive noise, the coherence has unit value when the noise is zero and gets smaller as the noise increases. In order to get the best possible estimate of the transfer function, we should try to work in the absence of noise, so *transfer function measurements should be made with good coherence*.

Frequency analysers capable of measuring transfer functions will be able to display coherence; you should *always* check the coherence associated with "twin channel" measurements. Practically, keeping $\gamma^2 \gg 0.9$ will ensure good results.

In summary: Low coherence implies statistical independence, which could be the result of:

1. Non-linear relationship (Perhaps the relationship is non-linear or perhaps there is no relationship at all.)
2. Presence of dependent noise.
    1. Sampling discretizing the signal in time
    2. Quantization (discretizing the signal in magnitude)

SUMMARY

Digitisation is the first and essential step that needs to be taken to handle analogue signals in a time-discrete and magnitude-discrete manner, i.e. the so-called digital manner. The most straightforward way to do this is to utilise multiple magnitudes with even steps and sampling points at even time-intervals. Fourier transform divulges the relationships between the time domain and the frequency domain representations of signals sampled at even points in time and in uniform frequency bins. This chapter outlined these essential concepts, which form the foundation of digital signal processing.

BIBLIOGRAPHY AND EXTENDED READING

Hayes, M. H. (2008) *Statistical Digital Signal Processing and Modelling*, Wiley.
Haykin, S. and Van Veen, B. (2002) *Signals and Systems*, Second Edition, John Wiley & Sons.
Oppenheim, A. V. and Schafer R. W. (2013) *Discrete-Time Signal Processing: Pearson New International Edition*, Pearson.
Oppenheim, A. V. and Willsky, A. S. (2013) *Signal and System: Pearson New International Edition*, Second Edition, Pearson.

EXPLORATION

1. Download the file *fourier.zip*, and unpack it in working directory of MATLAB. This zipped file contains a readme file, some short audio clips, and heavily commented example codes. These include the ones mentioned in chapter, which you may wish to try and run them as you read, and those for further exploration below.
2. Some suggested further explorations are listed below.
   - The MATLAB script *basescript.m* creates a 400 Hz sine wave signal, displays the signal and its spectrum, and plays the signal through a sound card. Run the program without changing it, and observe the spectrum produced and the sound output. Change the input frequency to 10000 Hz (line 20). Observe and explain what you see and hear.
   - The MATLAB script *aliasing.m* demonstrates the input and output relationship for a time sampling system. It generates a series of sine waves at a sampling frequency of 11025 Hz and plots output frequency versus input frequency. For each sine wave, the spectrum is found, and from this the output frequency is obtained (the program searches for the location of the spike in the spectrum). For those unfamiliar with MATLAB, it is well worth examining the script as a programming reference.
   - The following audio files are given in *fourier.zip*:
     yes16384.wav, esy.wav, and *yes4096.wav*
     Calculate and plot the spectrum for each of the signals, choosing appropriate windows in each case.

- Two wav files are given:
  signal_1.wav and *signal1and2.wav*
  Consider x to be the signal contained in the first wav file, and z the signal contained in the second. In this case, the relationship between the two signals was:
  z = x ⊗ y
  where ⊗ denotes convolution. Using Fourier techniques, recover the unknown time signal y.
- Two wav files are given:
  sine1.wav and *sine2.wav*
  Read these wav files into MATLAB and see how long it takes to carry out a Fourier transform (e.g. *clear all;y=wavread('sine2');tic;fft(y);toc*) Examine the input signals; they are very similar, but why the timing differences?
- The MATLAB function XCORR() calculates the auto and cross covariance for different signals. It is in the signal processing toolbox. Its use can be found in the example script *autocovariance.m*.
- Calculate the auto-correlation of the noise signal x explicitly using the fft() and ifft() functions following the summary shown in Table 2.1. Confirm that you get a similar shaped graph to using the MATLAB XCORR() function. Then look at some individual values and evaluate whether you get the same numerical answers. Consider why you might get different numerical answers, even though the graphs look similar. Provide a quick explanation.

3 DSP in Acoustical Transfer Function Measurements

Most common acoustical measurements are related to the frequency responses, phase response, or the complete transfer function of a system or an acoustics transducer. In this chapter, linear transfer function measurements will be studied using different methods. Certainly the scope of measurements of acoustic parameters and audio signals are much broader than these; some other measurements will be discussed in the subsequent chapters.

In many acoustics and audio applications, linear and time invariant (LTI) systems are desirable to handle signals. The LTI systems that process the signals are one of the major foci of this book. First and foremost, how do we describe or characterise such systems? This has been generally discussed in the previous two chapters: Impulse responses in the time domain or transfer functions in the frequency domain describe the behavior of an LTI system. But how do we measure impulse responses and/or transfer functions? What are the concerns and constraints in real-world acoustic and audio system measurements? How is DSP used to help with measurements? These are the questions to be addressed in this chapter.

There are three major challenges in acoustic and audio system measurements, namely signal-to-noise ratio, non-linearity, and time variance. These three challenges are all due to the fact that the system under investigation deviates from idealised LTI models. There are certainly some other issues and constraints.

3.1 ACOUSTICAL TRANSFER FUNCTION MEASUREMENT PROBLEMS

In theory, transfer function as a complete description of the input-output mapping relationship of an LTI system is a universal definition, regardless of the types of dynamic systems. The generally principle of the measurement of transfer functions, or their time domain representation impulse responses, are based on the fundamental relationships of LTI systems discussed in Chapter 1. In theory, any broadband stimulus covering the full range in the spectrum of interest with signal energy in all frequencies within that range can be used as stimulus. To ensure the "frequency fullness" requirement, white noise and pink noise are often used traditionally. A white noise has an advantage of having a completely flat spectrum, and represents thermal noise found in many electronic circuits; therefore, the signal-to-noise ratio is generally constant when measuring an electronic system. The spectrum of a pink noise is close to music or speech signals. Although environmental and ambient noises often show even more energy in the lower end of their spectra when compared with a pink noise, the use of pink noise in room acoustical transfer function measurement still offers some mitigation of the unevenness of the signal-to-noise ratio. The use of any broadband random noise means that the stimuli need to be either applied for a prolonged period of time or repeated to allow for statistical analysis. In addition, the random noise needs to be monitored continuously. In the presence of background noise, the required measurement duration or repetitions become even longer or more. This can become a problem in some scenarios.

Time variance, though we often try to avoid modelling it, is prevalent in room acoustics and electro-acoustic transducers. Sound transfer characteristics in enclosures (rooms) are fairly linear but, unfortunately, are also time variant due to temperature fluctuation and airflow. A loudspeaker is time variant (also shows non-linearity) due to the heat produced by the voice coil. Prolonged measurement means that the system may not satisfy the time invariant assumption during the measurement, which can cause measurement errors.

Although random noises are still used as stimuli in the measurements of acoustic properties and audio systems, many modern computerised measurement systems apply more effective deterministic measurement stimuli, such as maximum length sequence (MLS) and swept sine signals. There are a few particular problems or concerns in acoustical transfer function measurements to consider when developing, choosing, and applying a specific measurement method:

1. An acceptable measurement duration that satisfies the application needs and ensures the avoidance of time variance artefacts.
2. Low intrinsic noise susceptibility to avoid the need of too many repeated measurements and averaging.
3. A low crest factor of testing stimuli so that the amplifier and transducer can effectively feed a sufficient amount of energy into the system under testing without causing clipping distortions.
4. Intrinsically robust against time variance.
5. The ability to work with systems with some weak non-linearity.

3.2   TRANSFER FUNCTION MEASUREMENT USING MLS

Transfer measurement using noise works in general, but it became extraordinary useful when it was carried out with MLSs (pseudo-random noise) instead of random noise. Maximum length measurement systems enable the fast and accurate measurement of LTI systems, including acoustics systems. They have much greater immunity to noise than other transfer function measurement systems.

MLS measurement systems use very similar theories to those used by dual channel FFT analysers; MLS is a particular ruthless and efficient implementation of the method. Consider a noise signal $x(t)$ be applied to a linear time invariant acoustic system with an impulse response $h(t)$. The output from the measurement is a signal $y(t)$.

The time and frequency input/output relationships can be written loosely as:

$$y(t) = x(t) * h(t)$$
$$Y(f) = X(f)H(f)$$

(3.1)

where * denotes convolution and capital letters denote the frequency spectrum. Any reasonable dual channel FFT analyser will carry out measurement of H and h using cross-spectra techniques. While this has some noise immunity, it is not a quick process as averaging has to be undertaken. This is because the noise input, $x(t)$, ideally should be measured over an infinite amount of time to account for the natural fluctuations that any random noise signal has. Measurement has to be for an infinite time to gain population statistics rather than best estimates. Typically, 1024 averages give an accuracy of about 3%. Using a maximum length sequence noise source removes the need for averaging in many cases.

3.2.1 Maximum Length Sequences (MLSs)

The MLS has many properties similar to white noise, so the cross correlation techniques present in Chapter 2 can be used. MLSs are binary and bi-polar sequences, a string of +1 and −1. If an MLS is of degree n and the sequence is N bits long, then $N = 2^n - 1$, e.g. if $n = 12$, then the sequence is 4095 bits long. A length that is not 2^n is awkward because we cannot apply standard FFT algorithms. This problem has been overcome, however, by using a Fast Hadamard Transform (see extended reading for more details).

MLSs are generated from linear feedback shift registers constructed from a primitive polynomial. Consider constructing an MLS sequence of length $N = 2^4 - 1 = 15$. A primitive polynomial of length $n = 4$ is:

$$b(x) = x^4 + x^3 + 1$$

(3.2)

Primitive polynomials of other lengths can be found from tables in books or the internet (a table is given later). The corresponding shift register to generate the MLS sequence is (if typical logic gates are used as illustrated in Figure 3.1, then the zeros need to mapped on to −1 to make the sequence bi-polar):

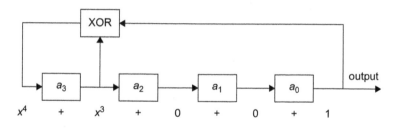

FIGURE 3.1 Illustration of MLS generation

Some starting conditions are given to the gates a_n, and then an m-sequence is generated.

(XOR is equivalent to addition modulo 2).

a_3	a_2	a_1	a_0	
1	0	0	0	
1	1	0	0	
1	1	1	0	
1	1	1	1	
0	1	1	1	
1	0	1	1	
0	1	0	1	
1	0	1	0	
1	1	0	1	
0	1	1	0	
0	0	1	1	
1	0	0	1	
0	1	0	0	
0	0	1	0	
0	0	0	1	
1	0	0	0	(Starts to repeat)

Primitive polynomials are studied in number, coding, and finite filed theories. The table below gives polynomials of degrees up to 16.

$$x^2 + x + 1$$
$$x^3 + x + 1$$
$$x^4 + x + 1$$
$$x^5 + x^2 + 1$$
$$x^6 + x + 1$$
$$x^7 + x + 1$$
$$x^8 + x^4 + x^3 + x^2 + 1$$
$$x^9 + x^4 + 1$$
$$x^{10} + x^3 + 1$$
$$x^{11} + x^2 + 1$$
$$x^{12} + x^6 + x^4 + x + 1$$
$$x^{13} + x^4 + x^3 + x + 1$$
$$x^{14} + x^5 + x^3 + x + 1$$
$$x^{15} + x + 1$$
$$x^{16} + x^5 + x^3 + x^2 + 1$$

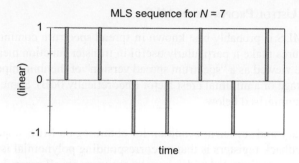

FIGURE 3.2 An illustration of an MLS sequence $N = 7$

Figure 3.2 illustrates an MLS sequence with $N = 7$. In software, it is easier to generate this using a modulo 2 sum:

$$a_n = \left(\sum_{i=1}^{n} b_i a_{n-i} \right) \mod 2 \tag{3.3}$$

Example

This is what is implemented in the script below:

```
%Code to generate an MLS sequence of order 10
close all; clear all
n=10;        %order of mls sequence
N=2^n-1;     %length of sequence
b = ([1,0,0,0,0,0,0,1,0,0,1])'; %taps
a = zeros(n,1);
a(1) = 1;    %Set latches with some starting values
for m=n+1:N
        %Sum of eqn 28, one state at a time modulo 2
        a(m) = mod( sum( flipud(a(m-n:m-1)).*b(2:n+1) ),2 );
end
c = 2*a-1;             %The best way to check if the mls sequence in
a is correct is
C = fft(c);            % to plot the autocorrelation
sa = round(real(ifft(C.*conj(C))));%all the sidebands should be
-1
plot(sa)
```

3.2.2 Some Useful Properties of MLS

Although the MLS is probably best known in spread spectrum communications, its interesting features make it particularly useful in transfer function measurement. In fact, it might be viewed as a "spectrum spread version" of the unit impulse function, with the advantage of a minimal crest factor (theoretically 0dB). Some properties of the MLS are summarised below:

1. A necessary and sufficient condition for obtaining an MLS sequence from linear feedback registers is that its corresponding polynomial is primitive. Registers can be initiated with any values, except for all zeros.
2. An MLS is periodic with a period of its length.
3. An MLS has a flat spectrum, i.e. equal signal power in every single frequency bin.
4. An MLS of length $2^n - 1$ contains 2^{n-1} "ones" and $2^{n-1} - 1$ "minus ones." If the sequence is reasonably long (as is often the case), it becomes very close to zero mean.
5. The circular auto-correlation of an MLS is (closely approximates, when the sequence is long) a unit impulse (Kronecker delta function):

$$Rxx = \begin{cases} 1, & n = 0 \\ \dfrac{1}{N}, & 0 < n < N \end{cases} \tag{3.4}$$

This property is used to measure the impulse response of an LTI system. Apparently, if an MLS is used to excite the LTI system, the impulse response of the LTI system can be obtained from the circular cross correlation of the system output with the MLS excitation.

3.2.3 Measure Once

It is only necessary to measure one complete period of the MLS sequence to determine the transfer function. As the MLS sequence is periodic, one complete period gives population statistics. This is in contrast to the white noise case, where to get population statistics, we had to average for an infinite amount of time. Consequently, MLS measurement is quicker than using white noise. Note that the MLS only has perfect auto-correlation properties when it is periodic, so in a measurement system, it is necessary to play two sequences: one to excite the filter being measured, and the second to actually make the calculations from.

3.2.4 No Truncation Errors

FFT analysers will calculate the cross correlation via Fourier techniques. This leads to the need for windowing the signals to prevent truncation errors. As the MLS sequence is periodic, no truncation errors result provided the calculation is carried out over exactly one period.

3.2.5 CREST FACTOR

The crest factor (peak/rms) is 1, which means that the signal uses the maximum available headroom in making measurements. This provides for a very high signa-to-noise ratio.

Imagine you are trying to record onto tape a piece of classical music and a piece of modern dance music. The dance music is highly compressed and so has a small crest factor. The signal spends most of its time near its maximum level. Consequently, when recording onto tape, the dance music is nearly always exploiting the maximum signal-to-noise ratio of the system. With most classical music, however, there is a wide dynamic range. You set the maximum level for the record, but then the quiet bits slip into the noise floor of the tape. Dance music has a small crest factor—it is like MLS—whereas random noise is more like the classical music case with a high crest factor.

Note, however, that there are suspicions that this small crest factor can cause problems in some transducers, leading to non-linearities and increased noise.

3.3 TRANSFER FUNCTION MEASUREMENT USING SWEPT SINE WAVES

MLS does not deal with distortion artefacts very well. The distortion appears as spikes in the impulse response leading to inaccuracies in the transfer function measurement. Swept sine waves have received increasing interest in recent years because they enable distortion artefacts to be removed. This enables us to get to the underlying linear process of a system. However, if the system is truly non-linear, shouldn't we be measuring that? Well, not always. If a system is being used to measure large rooms, then air currents in the room can cause the impulse response to vary with time; with MLS, this will produce distortion in the measured impulse responses. In this case, we want to get to the underlying linear response of the room without the time variance, so we can get to parameters such as reverberation time. Swept sine waves offer the possibilities of doing this.

3.3.1 MATCHED FILTERING

Consider a swept sine wave:

$$x(t) = \sin\left(\left[\frac{f_{end} - f_{start}}{2T}\right]t + f_{start}\right) \qquad (3.5)$$

where f_{start} and f_{end} are the starting and ending frequencies, and T is the total time for the sweep. If we take a time-reversed function of this, we get:

$$x_r(t) = \sin\left(\left[\frac{f_{start} - f_{end}}{2T}\right]t + f_{end}\right) \qquad (3.6)$$

Now the process of the measurement is this. We send out the swept sine wave $x(t)$ through a filter block $h(t)$. As we know, the result of this is:

$$y(t) = x(t) \otimes h(t) \qquad (3.7)$$

We then convolve the received signal with the time reversed input signal:

$$y(t) \otimes x_r(t) = \left[x(t) \otimes h(t)\right] \otimes x_r(t) = \left[x(t) \otimes x_r(t)\right] \otimes h(t) = h(t) \qquad (3.8)$$

where we have been able to change the order of the convolutions because they are simple multiply and add functions. There is an implicit assumption that $x(t) \bowtie x_r(t) = 1$. This is true of a sine sweep signal, and hence the name "matched filter" because the reversed swept sine is a matched filter of the original sine wave. However, while simply reversing the phase of the sine sweep produces a linear phase response, the magnitude of the impulse response is not flat. Consequently, we must divide the measured transfer function $H(f)$ by $X(f)X_r(f)$ to normalise.

Example

The measurement process is shown in the MATLAB® script *sweptsine_basescript.m*.

Proving that the reversed swept sine wave is a matched filter for the swept sine wave is easy numerically (see the MATLAB scripts), but doing it analytically is more tricky because the Fourier transform of a swept sine wave includes horrible unanalytical results (error functions). In practice, designing the inverse of a discrete time filter is actually not difficult if you follow the prescription in the following paper: O. Kirkeby, P.A. Nelson, H. Hamada, and F. Orduna-Bustamante, "Fast Deconvolution of Multichannel Systems Using Regularization," *IEEE Transactions on Speech and Audio Processing*, 1998, Vol. 6, No. 2, pp.189–194.

The swept sine wave system is good because inherently it has a low crest factor (the swept sine wave is a good test signal in this respect). Care has to be taken to ensure the sweep is not too fast; otherwise, the transfer function measurement becomes inaccurate, especially the phase. When noise is present, averaging is necessary to remove the noise:

$$y(t) \otimes x_r(t) = \left[x(t) \otimes \left(h(t) + n(t)\right)\right] \otimes x_r(t)$$

$$= \left[x(t) \otimes x_r(t)\right] \otimes h(t) + \left[x(t) \otimes n(t)\right] \otimes x_r(t) = h(t) + n(t) \qquad (3.9)$$

The system does not have such good noise immunity as MLS, but the noise can be reduced by averaging. Provided the noise is zero mean, it should gradually disappear. However, the swept sine wave system is good at dealing with distortion.

Distortion generates harmonics of the fundamental excitation, which is indicated in Figure 3.3. Most distortion creates higher frequency harmonics (an exception being quantisation noise, but that is unusual). As the diagram shows, this will generate components above the original swept sine curve, which, therefore, come out as earlier times in the match filter. The components we want, on the other hand, occur later in time, because the true impulse response produces stuff that is delayed in time but not shifted in frequency. Consequently, for a distorted signal, we see artefacts in the measured impulse response appearing earlier in time. These can be removed by simply excluding them by time gating and Fourier transforming the rest of the impulse response.

Because the sine sweep is a transient signal naturally, the start and end of the measurement period are not completely accurate. One solution is to sweep over a slightly wider bandwidth than required, and then disregard energy outside this bandwidth.

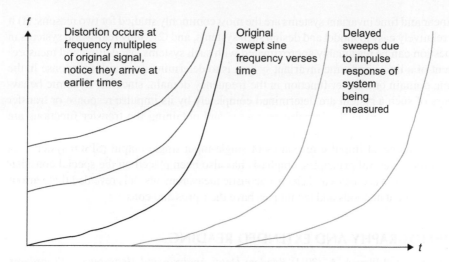

FIGURE 3.3 Illustration of distortions generated

SUMMARY

MLS	Random Noise	Swept Sine Wave
Used for linear, time invariant (LTI) systems	Used for linear, time invariant (LTI) systems	Best for linear, time invariant (LTI) systems, but can be used in time variant cases with small variance
Uses cross correlation to provide noise immunity	Uses cross correlation to provide noise immunity	Best in noise-free cases
Averaging often not needed	Averaging always needed	Averaging needed with external noise
High crest factor: good S/N, can cause problems	Lower crest factor, less S/N	Fairly high crest factor: good S/N
No windowing used	Windowing and overlap required	Used for cases where distortion is being eliminated.
If the MLS sequence is not long enough, the end of the LTI system impulse response wraps around to the beginning.	If the noise is not long enough, the end of the LTI system impulse response wraps around to the beginning.	If the sweep is not long enough, the end of the LTI system is lost.
To ensure no noise interference on input signal, output is correlated with internal input signal within computer.	Two transducers can be used, i.e. input signal can contain noise interference.	To ensure matched filter approach, output is convolved with internal reversed input signal within computer.
Distortion appears as spikes in the impulse response	Distortion appears as spikes in the impulse response	Uses matched filters and time windowing to exclude higher harmonics

Linear and time invariant systems are the most commonly studied for two reasons: (1) It is relatively easy to model and design such systems, and (2) Many real-world systems in question can be reasonably approximated with such systems. Modelling and measurement of a linear and time invariant system is to determine its impulse response in the time domain or transfer function in the frequency domain, since the dynamic behaviours of such a system are determined completely by its impulse response or transfer function. Acquiring the impulse responses and obtaining the transfer functions are deemed "equivalent," as they are Fourier transform pairs. This chapter focused on the measurements of impulse responses of single-input-single-output (SISO) systems. In addition to general principles, emphasis has also been placed on the special considerations in room acoustics and electro-acoustic measurements. It is revealed that various measurement methods and techniques have their pros and cons.

BIBLIOGRAPHY AND EXTENDED READING

Bendat, J. and Piersol, A. (2011) *Random Data: Analysis and Measurement Procedures*, Fourth Edition. John Wiley & Sons.
    A very useful general text on transfer function measurements; not easily accessible but the most authoritative.
Bradley, J. S. "Optimizing the Decay Range in Room Acoustics Measurements using Maximum Length Sequence Techniques," *Journal of the Audio Engineering Society*, 1996, Vol. 44, No. 4, pp. 266–273.
    Interesting sections on distortion artifacts and their influence on RT measurement.
Farina, A. "Simultaneous Measurement of Impulse Response and Distortion with a Swept-sine Technique," 108th Convention of the Audio Engineering Society, 5093 (D - 4), (2000).
Rife, D. D., Vanderkooy, J. "Transfer-function Measurement with Maximum-length Sequences," *Journal of the Audio Engineering Society*, 1989, Vol. 37, No. 6, pp. 419–444.
    The original paper setting out the basis for the commercial MLS system MLSSA.
Stan, G. B., Embrechts, J. J., and Archambeau, D. "Comparison of Different Impulse Response Measurement Techniques," *Journal of the Audio Engineering Society*, 2002, Vol. 50, No. 4, pp. 249–262.
    A nice comparison of sine sweep and MLS measurement methods.
Svensson, P., Nielsen, J. L. "Errors in MLS Measurements Caused by Time Variance in Acoustic Systems," *Journal of the Audio Engineering Society*, 1999, Vol. 47, No. 11, pp. 907–927.
Vanderkooy, J. "Aspects of MLS Measuring Systems," *Journal of the Audio Engineering Society*, 1994, Vol. 42, No. 4, pp. 219–241.
Vorlander, M., Kob, M. "Practical Aspects of MLS Measurements in Building Acoustics," *Applied Acoustics*, 1997, Vol. 52, No.3-4, pp. 239–258.

EXPLORATION AND MINI PROJECT

EXPLORATION

1. Download the zipped file *measurement.zip* and unpack it in working directory of MATLAB. This zipped file contains a readme.txt and heavily commented skeleton codes as examples and demonstrations for the measurement methods discussed in this chapter. These include methods that use white (or broadband random) noise, sine sweep, or MLS as testing stimuli.

2. Assume that a room response to be measured can be approximated by a simple Chebyshev filter below (note that this is a convenient choice for exploring the measurement technique, but the filter is not a true representation in many aspects.)

```
[b,a] = cheby1(8,1,0.4);
y = filter(b,a,x);
```

Use three different stimuli, namely white noise, MLS, and sine sweep, to measure the above acoustic transfer function of the above "room," and compare the performances of these measurement techniques in terms of signal-to-noise ratio, measurement duration, noise susceptibility, and the ability to handle non-linearity.

Some additional suggestions: As an effective way to get the investigation and exploration organised, you can consider the following three statements. They can be either true or false, or partly true. Give your judgments and support your arguments by giving simulation evidence.

Statement 1: Measurements in a Background-Noise-Free Environment

"In a noise-free environment, measurements using MLS and white noise as excitation require averaging to achieve good accuracy, whereas measurements using swept sine do not. To achieve similar frequency response measurement accuracy, swept sine is the fastest technique, followed by MLS, then white noise. MLS is more accurate than white noise because it is ideally decorrelated from itself."

Statement 2: Measurements in the Presence of Background Noise

"With background noise present, it is often necessary to carry out averaging to get an accurate estimation of the room frequency response. White noise excitation has better immunity to background noise than MLS because of its higher crest factor. Swept sine technique is not as good at dealing with background noise as MLS. Impulsive background noise (such as coughing) produces clear artefacts in the room impulse response for all methods, but swept sine technique can remove them completely."

Statement 3: Loudspeaker Distortion

"When loudspeaker distortion occurs during an MLS measurement, distinct spikes appear in the impulse response for each distortion harmonic; these cannot be removed by averaging. Swept sine technique allows the distortion artefacts to be removed from the impulse response so the underlying linear system may be measured."

Hints

- When considering measurement time, this is the time to perform a real measurement, not the time it takes MATLAB to do the calculations to simulate it.
- MLS assumes periodic excitement but the filter initiates in a silent state. Is the first period equivalent to one period of a periodic signal? If not, what can you do to fix it?

- Background noise is sound picked up by the microphone that is not caused by the test signal. This is usually considered to be white noise added to the measured signal after it has been filtered by the room.
- The sound of a cough is provided in the file *cough.wav*. Try adding it to the swept sine measurement approximately halfway through its duration and plot spectrograms to picture what is going on.

Starting Points

- The script *mls_vs_noise.m* passes an MLS signal of length 1023 bits (contained in *mls1023.wav*) through an LTI system, so it is a good place to start in setting up your measurement system. If you want other length MLS sequences, they can be created using *mls_generator.m*.

- The script *impulse_averages.m* might be a useful starting point for considering the effects of averaging. When you set up the MLS and white noise methods correctly, the graph shows the type of results you can expect when there is no background noise (NB the filter measured here is slightly different). Notice the very high measurement signal-noise ratio. If you don't

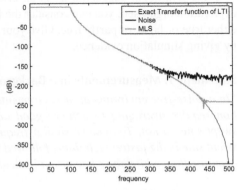

achieve such large S/N ratios in the noise free case, something is wrong with your code.
- The script *swept_basescript_conv.m* implements swept sine measurement.
- *Loudspeaker.m* contains a crude model of loudspeaker distortion.

Mini Project Ideas

In the suggested exploration, details and codes for the commonly used methods for impulse response measurements have been experimented with. These include the core signal processing part of a measurement system for the transfer function or impulse response of LTI systems. Late versions of MATLAB support easy-to-use audio I/O via USB audio interfaces. The project idea is to develop a fully functioning LTI measurement system on the MATLAB platform using an external USB audio interface.

It is worth noting that for reliable measurement and a low level of system noise, a high quality external audio interface is recommended. Of course, internal soundcards will work, but signals are more likely to be contaminated by noises coupled from buses and power suppliers.

4 Digital Filters and z-Transform

4.1 GENERAL INTRODUCTION TO DIGITAL FILTERS

As discussed earlier in this book, a system manipulates or processes the signals passing through it. Filters are purposely designed and used to manipulate the signals. Filter terminology stems from the traditional circuits or systems in which certain unwanted signal components need to be removed or reduced. For example, a guitar amplifier may have a circuit to remove the mains induced 50 or 60 Hz hum noise. A radio receiver may have tone control knobs to set the amount of low-frequency or high-frequency components to be attenuated to offer preferred tonal balance by the listeners. These are essentially the early form of filters: a circuit that selectively passes through certain frequency components and filters out other frequency components; hence the name. Low pass, high pass, band pass, and band stop filters are all terminologies from traditional filters whose purpose are inlay frequency component selection. Modern filters and especially digital filters certainly can do much more than traditional ones can do. The scope of filters has gone far beyond the traditional sense. Since a system manipulates signals, and a filter is designed and used to manipulate the signals, in a broad sense any system can be viewed as a filter. In fact, systems are often modeled as a filter. The task of designing a filter is the implementation or synthesis of a system that satisfies certain design criteria. The difference between a system and a filter is that a system may or may not be a product of an intentional effort, but a filter typically is. For example, a room modifies sound by adding reflections and reverberation. A room is often viewed as a sound transmission system; we can model it using a filter. The straightforward way to completely model the point-to-point sound transmission characteristics in a room is to capture its impulse response, but to implement such a notoriously long (infinite in theory) and irregular filter is virtually impossible. To emulate acoustics of rooms, various artificial reverberators are designed, taking into account the ease of implementation. There are also filters designed to equalise or compensate for certain undesirable room responses. These are typical examples to differentiate between a system and a filter.

Since filters are produced by design, and the purpose of design is to implement and apply it to process the signals, we will see later that much of filter realisation takes into account various constraints, ease of implementation, and cost effectiveness.

Digital filters are known to be flexible and can be cheap to implement with today's technologies. To highlight these particular advantages, a historical view of various electronic filters would be useful. The earlier filters are known as passive filters. They utilise only three types of components, namely resistors, capacitors, and inductors, and must have at least one capacitor or one inductor to selectively pass through wanted frequency components of signals. To achieve frequency selection, some form of memories must be used. In passive filters, energy storage devices, capacitors, and inductors serve as memories. The lower the frequencies these filters are to function, larger capacitance and inductance are needed, making physical devices bigger, heavier, and more expensive. Capacitors and inductors of extremely high values can be difficult or even impossible to manufacture. In digital filters, the memories are achieved using the logical status kept by, say, a D flip-flop instead of energy. The status can be kept for as long as is needed. Therefore, digital filters can be extended easily to very low-frequency applications. However, at the high end of the spectrum, the frequency bandwidth is limited by the digital system clock. As a result, for very high-frequency applications, passive analogue still find their applications.

Digital filters can be implemented using hardware or software. Of course, software must run on certain hardware platforms, and DSP hardware is mostly programmable. For hardware implementation, digital filters are realised using only three types of building blocks: delay units, summers, and multipliers. Mathematically, digital filters can be described by linear difference equations. Therefore, software implementation essentially is the algorithms that solve these equations. The logic and arithmetic unit (ALU) in the central processing unit (CPU) of a binary computers are designed for logic and addition operations, but not multiplication. Multiplication operations are done by multiple summations. This makes general purpose CPU less effective for multiplication applications, such as DSP, if real-time performance is critical. Hardware DSP platforms are designed with built-in hardware multipliers to improve performance in DSP tasks.

While digital filters are flexible due to a number of reasons, there is one most important reason. The design target of a digital filter is to find a suitable set of difference equations. These can be translated into computer algorithms and codes. If the design is not intended for real-time application, algorithms should have no constraint from physical reality. With some data buffering, filter algorithms can even implement filters that are not physically possible, e.g. non-causal filters in which the output can take place before the excitation is applied. Software implementation of filter algorithms and simulation make no major difference. One can implement filters up to very high frequency without having to have a computer operate at half of the frequency.

Digital filter is a huge area of study. But in most cases, the implementations of linear invariant systems are considered for their simplicity and the availability of well-studied design techniques. To handle certain non-linearity and time variance, adaptive filters with a linear time invariant core can adapt the parameters to environment, and behave in a nonlinear and time variant way. In this chapter, we will focus on linear digital filters and in the next chapter, adaptive filters.

To realise an LTI system, one may start from the general form of its difference equation. In theory, there can be many possible structures. However, this is a huge knowledge base of using a few generally regular structures to implement the difference equations. Indeed, the study into digital filters follows the route of studying their structures, and the filters are classified by their structures.

Finite impulse response (FIR) filters and infinite impulse response (IIR) filters are the two general classes. The former, as their name suggested, have finite length impulse response. They are implementing with a feed forward structure. With a limited number of storage units and a feed forward structure, their impulse responses have limited lengths. The latter, IIR filters, attempt to reuse the system resources for an infinite number of times by looping the output back to its input and, hence, having impulse responses of infinite lengths. For each of these two general classes, there are several commonly used subclasses often called types.

This chapter highlights the essential concepts, and then presents typical acoustics and audio related application and a few important legendary audio filters. These will be detailed in the next few sections.

4.2 FINITE IMPULSE RESPONSE (FIR) FILTERS

Consider the LTI system as depicted in Figure 4.1, in which $x(t)$, $y(t)$, and $h(t)$ are the input, output, and impulse response, respectively.

Recall the input the relationships between input and output,

$$y(t) = x(t) \otimes h(t) \tag{4.1}$$

If $X(j\omega)$, $Y(j\omega)$, and $H(j\omega)$ are the Fourier transform of $x(t)$, $y(t)$, and $h(t)$, respectively, then

$$Y(j\omega) = X(j\omega)H(j\omega) \tag{4.1'}$$

If a discrete time system is considered

$$y[n] = x[n] \otimes h[n] \tag{4.1''}$$

And as will be discussed later, in the z-transform domain, which is the Laplace transform counterpart for the discrete time signals,

$$Y(z) = H(z)X(z) \tag{4.1'''}$$

This ought to be familiar as it is the equation we had for an LTI system. So all those illustrations seen before of impulse responses, transfer functions, and input and output relationships are all relevant. You might want to run the MATLAB® file *impulse.m* to appreciate these.

In short, the FIR filtering process is convolution in the time domain and multiplication in the frequency domain.

Consider the convolution equation in more detail:

$$y_k = \sum_{j=0}^{N-1} b_j x_{k-j} \tag{4.2}$$

This represents what is happening in the discrete time domain. The set of numbers b are known as the filter coefficients or taps, and there are N taps in this case.

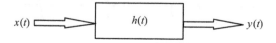

FIGURE 4.1 An LTI system

The MATLAB function to do this convolution is *conv()* (you could also adapt the function *filter()* to do this).

Run the MATLAB script *anFIR.m*. This takes a swept sine wave and passes it through a low pass filter. MATLAB Figure 2 shows how the input signal's frequency increases with time, and also how low frequencies are passed through the filter and high frequencies are attenuated. Examine the script to see how it works.

Question: The output signal (y) is longer than the input signal (x). Why is this?

Figure 4.2 illustrates the filtering process as a combination of multiplication (gains in triangles), delays (z^{-1} means delay one sample), and accumulation (summing). The power of this structure is that by changing the coefficients, b's, we can get a completely different filter (low pass, high pass, band pass, band stop, etc.) from the same circuit.

It is apparent that the structure shown in Figure 4.2 is a direct representation of Equation 4.2, hence the name of "direct form." A series of delay taps provides the memories needed while filtering, so this type of filter structure is also known as a tapped delay line. The tapped delay line intersects a number of weighting lines to give the summed output; as a result, such a structure is also called a transversal filter.

There are a number of interesting features of an FIR filter:

- Symmetrical FIR filters are automatically phase linear (phase is proportional to frequency).
  This is preferable for many audio applications and many other applications.
- FIR filters are completely stable.
  The output is the weighted sum of limited taps of delayed input. Bounded input guarantees bounded output. There are, therefore, very few design constraints, providing great flexibility.
- FIR filters can become too computer/component expensive and take too long to process for real-time applications when a large number of coefficients (taps) are needed to achieve design specifications.

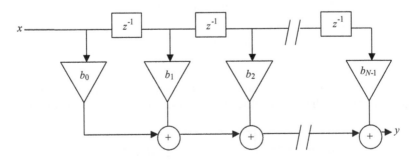

FIGURE 4.2   Direct form of FIR filter

4.3   z-TRANSFORM AND TRANSFER FUNCTION

Since the b_k coefficients represent a discrete impulse response, it should be possible to obtain the filter frequency response by using an appropriate transform to the frequency domain. In the Tutorial question 1, you actually do this using a Fourier transform. However, in the case of digital filters, the z-transform will turn out to be even more useful. (But the Fourier and z-transforms are closely related). The advantage of the z-transform is it readily enables the frequency response and stability of digital filters to be explored through the pole-zero representation.

The z-transform is the digital equivalent of the Laplace transform and is defined as follows:

$$X(z) = \sum_{n=0}^{\infty} x_n z^{-n} \tag{4.3}$$

Each element of the digitised sequence x_n is multiplied by the complex exponential z raised to a negative power corresponding to the position of x_n. Note that this is almost identical to the discrete Fourier transform (DFT). If we substitute $z = e^{j\Omega}$ into Equation 4.3, we get

$$X(\Omega) = \sum_{n=0}^{\infty} x_n e^{-j\Omega n} \tag{4.4}$$

which is a one-sided Fourier transform (one sided because n is never negative). Equation 4.3 is, in fact, a power series in z^{-1}:

$$X(z) = x_0 z^{-0} + x_1 z^{-1} + x_2 z^{-2} + \dots. \tag{4.5}$$

Consequently, z is often seen as a time shift operator. As we multiply by z^{-1} we go forward in time, i.e. the index of x the signal increases. The z-transform is generally applied term by term to the digital filter difference equation using the standard z-transforms.

$$x_n = X(z)$$

$$x_{n-1} = X(z)z^{-1} \tag{4.6}$$

$$\beta x_{n-\alpha} = \beta X(z)z^{-\alpha}$$

Having z-transformed each term of the filter difference equation, we obtain a new equation in z that gives the frequency behaviour of the filter. We will see two examples in a few subsections showing how we analyse filters using difference equations, z-transfer functions, and zero and pole plots.

Example

Three-point moving average filter

A moving average filter is an FIR filter where the coefficients are the same.

$$y_n = \frac{1}{N}\sum_{k=0}^{N} x_{n-k} = \frac{1}{N}\left[x_n + x_{n-1} + x_{n-2}\right] \tag{4.7}$$

This is the *difference equation,* which gives a description of the filter behaviour in discrete time. It is the most efficient way of implementing a filter with a small number of coefficients. To obtain the z equation (i.e. verses frequency), we z-transform each term according to the standard transforms given in Equation 4.6.

$$Y(z) = \frac{1}{N}\left[X(z) + X(z)z^{-1} + X(z)z^{-2}\right] \tag{4.8}$$

Then rearrange to get the filter response, which is given by the filter output $Y(z)$ divided by the input $X(z)$

$$\frac{Y(z)}{X(z)} = \frac{1}{N}\left[1 + z^{-1} + z^{-2}\right] \tag{4.9}$$

And the final step is to divide through by z^2 to obtain a rational polynomial in z.

$$\frac{Y(z)}{X(z)} = \frac{1}{N}\left[\frac{z^2 + z + 1}{z^2}\right] \tag{4.10}$$

The filter response is determined by the right-hand-side of the equation. The roots of the numerator govern the filter response minima (*zeros*) and the roots of the denominator govern the filter maxima (*poles*). The positions of the poles and zeros in the complex plane give the filter frequency response.

4.4   ZERO-POLE PLOTS

We are now ready to describe the filter response using a pole zero diagram. $Z = \exp(j\Omega)$ is a unit vector that describes a circle in the complex plane called the unit circle, Figure 4.3.

The distance around the circle represents increasing frequency from DC at 1 on the real axis and the Nyquist frequency at −1 on the real axis. The frequency at any point on the unit circle is therefore given by:

$$\omega = \frac{\theta.fs}{2\pi} \tag{4.11}$$

Just as an analogue filter has poles and zeros that govern the filter response in the s-plane, so does the digital filter in the z-plane. The response for a particular frequency is given by the ratio of the distance to the zero to the distance to the pole from the point on the unit circle representing that frequency.

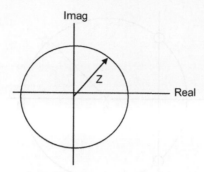

FIGURE 4.3   Unit vector and unit circle

Go back to the three-point moving average as given in Equation 4.10. The general rule is that the roots of the numerator determine the position of the zeros (at these points, the filter response is zero), and the roots of the denominator determine the position of the poles where the filter response is maximal.

In this example, the top has roots at

$$\frac{-1 \pm \sqrt{1-4}}{2} = -0.5 \pm j \frac{\sqrt{3}}{2} = -0.5 + 0.866j \text{ and } -0.5 - 0.866j$$

and the bottom has two roots at the origin. The zero-pole plot and magnitude response are given in Figure 4.4.

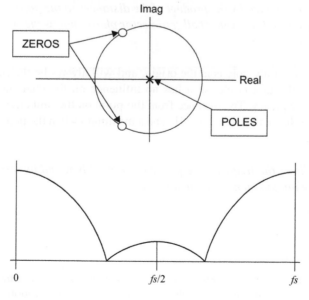

FIGURE 4.4   Zero-pole plot and frequency response of the three point moving average filter

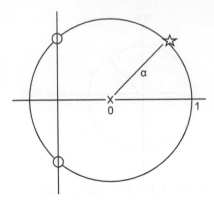

FIGURE 4.5   Illustration of visual estimation of frequency response from Zero-pole plot

This is the response for a low pass filter as expected (a moving average is a smoothing filter, so high frequency variation is removed and, hence, this is a low pass filter). Notice that the form of the filter response is reflected about $fs/2$, and since we are not interested in frequencies above $fs/2$, we can disregard this—these are the negative frequencies met before with the Fourier transform.

The frequency response can be visually estimated from the z-plane, as shown in Figure 4.5. The frequencies are traced out by following the top half of the unit circle. At an angle α radians, the frequency is given by $\alpha f_s/(2\pi)$, so in the above diagram, the frequency is about ¼ of the way to the Nyquist frequency. Then:

- *The filter magnitude response (gain) is given by the product of the distances to the zeros divided by the product of the distances to the poles.*
- *The phase equals the sum of all zero-vector phases minus the sum of all the pole-vector phases.*

In this case, the only pole is at the origin, and will always be the same distance to the point on the unit circle, so it has no influence on the changing magnitude response with frequency. The distance from the point on the unit circle to the zeros varies, however. It is a maximum at 0 Hz and a minimum when the point is on a zero, as was shown.

The z-plane allows the frequency response to be readily found by inspection (provided there are not too many poles and zeros).

Example

High Pass Filter

Suppose that instead of summing the discrete values, we subtract the previous value from the current value. We will then generate an output proportional to the difference between subsequent values of the digitised signal. This is effectively the

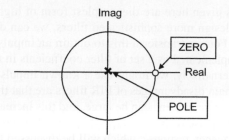

FIGURE 4.6   Zero-pole plot of a high pass filter example

well known differential pulse-code modulation, or DPCM. Consequently, this will be greater for high frequencies than for low frequencies and we will have a crude high pass filter.

We can write the difference equation as before:

$$y_n = \frac{1}{N}\left[x_n - x_{n-1}\right] \tag{4.12}$$

And applying the z-transform as before, we obtain

$$Y(z) = \frac{1}{N}\left[X(z) - X(z)z^{-1}\right] \tag{4.13}$$

And dividing through by $X(z)$ obtains the response

$$\frac{Y(z)}{X(z)} = 1 - z^{-1} \tag{4.14}$$

Multiplying top and bottom by z to obtain a rational polynomial gives

$$\frac{Y(z)}{X(z)} = \frac{z-1}{z} \tag{4.15}$$

The top of the equation has one root at $z = 1$ on the real axis and a pole at $z = 0$. The zero-pole plot is show in Figure 4.6 and the magnitude-frequency response in Figure 4.7.

And the filter response will be of the form shown in Figure 4.7.

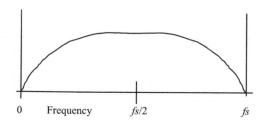

FIGURE 4.7   Magnitude-frequency response of a high pass filter example

The two examples given here are the simplest form of high and low pass FIR filter. If we want to design more sophisticated filters, we can define the frequency response and inverse Fourier transform this to obtain an impulse response that can be discretized and implemented as a set of filter coefficients in the general FIR difference equation. Alternatively, we might use a known impulse response from an analogue filter. The main disadvantages of FIR filters are that the number of coefficients required for accurate filtering can be large, and this increases the computation time. Shortening the impulse response to speed up the filter introduces "truncation" errors in the filter frequency response, which will be discussed later.

4.5   INFINITE IMPULSE RESPONSE (IIR) FILTERS

Infinite impulse response (IIR) filters have a recursive structure. One of the simplest examples is shown in Figure 4.8; it is a comb-filter based artificial reverberation unit. The recursive nature makes IIR filters ideal for repeated mechanisms such as reverberation. In theory, an IIR filter can model an infinitely long impulse response with only a finite number of coefficients.

Properties:

- Less circuitry than FIR filters.
- Harder to design than FIR filters.
- Not phase linear.
- Can be unstable if $h(t)$ has a gain > 1.
- Prone to round-off errors in the coefficient values as they are used many times.

The following equation (known as a difference equation) defines a very simple IIR filter. In this case, $h(t)$ is simply an attenuation factor of 0.5.

$$y_k = x_k + 0.5 * y_{k-1} \tag{4.16}$$

Run the MATLAB script *anIIR.m*, which implements Equation 4.16, and look at tutorial question 2. This is a pretty awful filter! We need a more complex difference equation, with many more terms, to get a decent filter:

$$y_k = \sum_{i=0}^{Np-1} b_i x_{k-i} - \sum_{j=1}^{N_0} a_j y_{k-j} \tag{4.17}$$

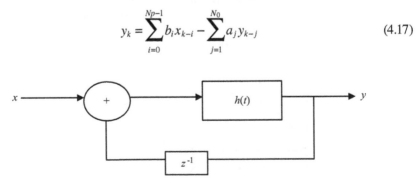

FIGURE 4.8   A simple IIR filter

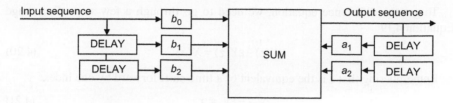

FIGURE 4.9 A Universal block diagram of IIR filters

The term *difference equation* crops up again; note this as we shall keep returning to it. If all the terms a_j are zero, we have an FIR filter (Equation 4.2). If any of the a_j terms are non-zero, this means past values of the output influence future values; we have recursion and an IIR filter.

This use of previous outputs to calculate the current output is known as *recursion* and it is an extremely efficient computational technique, resulting in this case in versatile digital filter designs that require few coefficients and, hence, minimal computation time. A block diagram of a more involved IIR filter is shown in Figure 4.9.

One way to understand these filters is to think of them as calculating their impulse response "as they go," rather than calculating the product of the signal and a set of fixed coefficients to obtain each output value as in the FIR filter. We will now see how we can use the z-transform and pole zero plots to design some real IIR filters.

The general z-domain transfer function for an IIR filter is

$$H(z) = \frac{b_0 + b_1 z^{-1} + b_2 z^{-2} + \ldots\ldots b_{m-1} z^{1-m}}{1 + a_1 z^{-1} + a_2 z^{-2} + \ldots + a_n z^{-n}} \tag{4.18}$$

Question: Can you derive this equation? Go into MATLAB and type "help filter()"; note this implements the same equation except that the indexing of a_n and b_n are a bit different.

The roots of this equation allow the filter poles to be placed anywhere in the z-plane, and, as such, it can be used as the basis for a variety of digital filters.

4.6 STABILITY

One of the problems with IIR filters (maybe the most important problem) is that they can become unstable. Consider a simple filter with a single pole at $z = \varepsilon$; by definition, this leads to:

$$H(z) = \frac{Y(z)}{X(z)} = \frac{1}{z - \varepsilon} \tag{4.19}$$

To get the difference equation, we need to go through a few steps. Rearrange Equation 4.19:

$$zY(z) - \varepsilon Y(z) = X(z) \tag{4.20}$$

Remembering that z is the equivalent of a time shift forward by one index:

$$y_{n+1} - \varepsilon y_n = x_n \tag{4.21}$$

which leads to the difference equation:

$$y_n = x_{n-1} + \varepsilon y_{n-1} \tag{4.22}$$

Alternatively, Equations 4.18 and 4.19 can be compared to identify the coefficients a and b, and the results substituted into Equation 4.17.

From inspection of Equation 4.22, it can be seen that the system can become unstable. Sceptical? Then try some simple MATLAB code:

```
eta=1.5;
n=100;
x=zeros(n,1);
x(2)=1;
y=zeros(n,1);
for j=2:n
        y(j)=eta*y(j-1)+x(j);
end
plot(y)
```

See what happens when you vary eta (ε). Which of the following statements is true?

- For an IIR filter to be stable, all the poles must lie within the unit circle.
- For an IIR filter to be stable, all the poles must lie outside the unit circle.
- For an IIR filter to be stable, all the zeros must lie within the unit circle.
- For an IIR filter to be stable, all the zeros must lie outside the unit circle.

4.7   BILINEAR IIR FILTERS (BILINS)

As the name implies, this design method involves using a pair of linear equations in z to derive the filter response. We begin with a difference equation of the form:

$$y_n = x_n + b_1 x_{n-1} - a_1 y_{n-1} \tag{4.23}$$

and carry out a z-transform as before and re-arrange:

$$Y(z) = X(z) + b_1 X(z) z^{-1} - a_1 Y(z) z^{-1} \tag{4.24}$$

$$Y(z) + a_1 Y(z) z^{-1} = X(z) + b_1 X(z) z^{-1} \tag{4.25}$$

$$Y(z)\left(1 + a_1 z^{-1}\right) = X(z)\left(1 + b_1 z^{-1}\right) \tag{4.26}$$

and finally

$$\frac{Y(z)}{X(z)} = \frac{\left(1 + b_1 z^{-1}\right)}{\left(1 + a_1 z^{-1}\right)} \tag{4.27}$$

This is an equation in z as before. To eliminate the z^{-1} terms, we multiply top and bottom by z to obtain:

$$\frac{Y(z)}{X(z)} = \frac{(z + b_1)}{(z + a_1)} \tag{4.28}$$

which has poles and zeros in the complex plane as before. The positions of the poles and zeros are governed by the values of a_1 and b_1 as before.

Example

Bilinear high pass filter

We can implement a high pass filter with a zero at DC (0 Hz) by making $a_1 = 0$ and $b_1 = -1$.

This will give the high pass filter that we obtained previously. We can also modify the filter characteristic by assigning a value to b_1 that has the effect of moving the pole away from the origin where it can have an effect on the filter behaviour, specifically, the pole radius controls the corner (approximately −3dB) frequency.

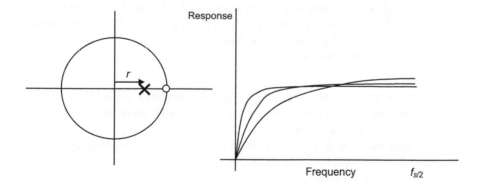

The three curves show the filter response for varying a_1. Notice that the filter response becomes steeper as a_1 approaches 1, and if it exceeds or equals 1, the filter will become unstable. The bilinear design method allows us to move the pole away from the origin; however, it cannot move away from the real axis. It is only really useful for understanding filtering and implementing the most basic possible filters. To place poles and zeros anywhere within the unit circle, we need to consider a higher order filter and use the biquad method.

4.8   BIQUADRATIC IIR FILTER DESIGN (BIQUADS)

In this case, we derive a quadratic z equation.

$$\frac{Y(z)}{X(z)} = \frac{\left(z^2 + b_1 z + b_2\right)}{\left(z^2 + a_1 z + a_2\right)} \qquad (4.29)$$

Just as before, the roots of the top of the equation give the zero positions and the roots of the bottom give the pole positions. The general design method is to decide where we want to put poles and zeros to obtain the required filter, and to calculate the a and b coefficients from the pole/zero co-ordinates in the z-plane. Poles and zeros in the z-plane always exist in pairs (complex conjugates) since they arise from the solution of a quadratic equation. (If they do not exist in complex conjugate pairs, the coefficients of the difference equation become complex, and so the filter cannot be made because we cannot have complex time variables.)

If we have a pole pair in the z-plane with coordinates $R_p +/- jI_p$, the a coefficients are given by

$$a_1 = -2R_p \quad \text{and} \quad a_2 = \left(R_p^2 + I_p^2\right)$$

Similarly, if we have a pair of zeros, the b coefficients are given by

$$b_1 = -2R_z \quad \text{and} \quad b_2 = \left(R_z^2 + I_z^2\right)$$

Having decided on the pole and zero positions, and calculated the a and b coefficients, we then inverse transform the z equation to get back to a difference equation in time, which can be coded into a digital filter.

Thus, the design procedure is:

1. Decide on pole and zero positions for the filter required. (This takes a certain amount of judgement and experience, although there are interactive tools to help. See the Study Guide).
2. Calculate the a and b coefficients from the pole and zero coordinates.
3. Inverse z-transform to obtain the difference equation.
4. Implement the difference equation as a digital filter.

Example

Band stop filter (notch filter)

This type of filter has a zero = response at a particular frequency and passes all other frequencies within the baseband. The pole zero plot is shown below.

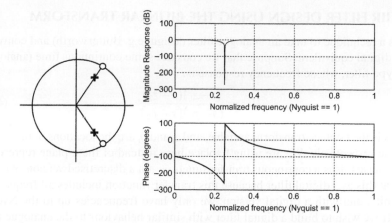

Remember that the frequency response is obtained by moving around the unit circle and is given by the distance to the zero divided by the distance to the pole from the point on the circle representing a particular frequency. The Q (or sharpness) of the filter is determined by the pole radius.

The difference equation is

$$y_n = x_n - [2\cos\Omega]x_{n-1} + x_{n-2} + [2r\cos\Omega]y_{n-1} - r^2 y_{n-2}$$

where $\Omega = \dfrac{2\pi f}{f_s}$ and f is the frequency to be filtered out.

Example

Band pass filter

The difference equation in this case is

$$y_n = x_n + x_{n-2} + [2r\cos\Omega]y_{n-1} - r^2 y_{n-2}$$

Using these design methods we can implement a range of useful filters to suit our requirements, but it is a bit tedious if you are not very experienced at pole-zero placement.

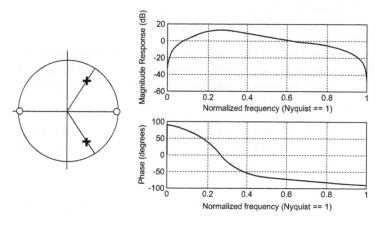

4.9 IIR FILTER DESIGN USING THE BILINEAR TRANSFORM

This is a technique to take an analogue filter design (e.g. Butterworth) and convert it into a digital equivalent. Suppose that we consider some continuous time (analogue) prototype filter, whose frequency response is:

$$H(s) = \frac{(s - z_1)(s - z_2)}{(s - p_1)(s - p_2)} \tag{4.30}$$

where s is the Laplace transform variable, and z_i and p_i are the locations of the ith zero and pole respectively. We are using Laplace here instead of the z-plane representation because this is a continuous transfer function, not a discretised version. We cannot treat this as a digital filter because this transfer function includes all frequencies to infinity, and with a digital version, we only have frequencies up to the Nyquist limit. If we wish to build a digital filter with similar behaviour to the analogue filter described above, we could take two approaches:

1. Acknowledge that the digital filter can copy the analogue prototype's frequency response only to the half-sample frequency. This is equivalent to the FIR transfer function design technique shown later on.
2. Attempt to "compress" the infinite frequency range of the analogue system into the finite range of the digital system. This is what we will do here.

To make the compression, we need some function that sweeps in magnitude from zero to infinity as the normalised frequency sweeps from 0 to half the sample frequency (or as the angle from the origin in Z to a point on the unit circle sweeps from 0 to π. One such function is the simple bilinear form:

$$B(z) = \frac{z - 1}{z + 1} \tag{4.31}$$

Now Equation 4.31 has a zero at 0 Hz and a pole at $f_s/2$. The response, calculated as a function of Ω ($=\omega T$) is:

$$F(\Omega) = \frac{e^{j\Omega} - 1}{e^{j\Omega} + 1} \tag{4.32}$$

which can be rearranged to give:

$$F(\Omega) = j \tan(\Omega/2) \tag{4.33}$$

$F(\Omega)$, as shown in Figure 4.10, is imaginary valued and its magnitude will increase from zero to infinity as Ω sweeps from zero to π; this is exactly the type of function we need to compress the infinite frequency range of a continuous filter into the finite frequency range of a digital system.

Notice that, whilst the compression is reasonably linear around dc (as the tangent function has unit gradient at $\Omega = 0$, the mapping is highly non-linear nearer to the half-sample frequency.

Therefore, if we substitute $F(\Omega)$ for s in the Laplace domain expression of any analogue filter, we shall obtain exactly the same values for the transfer function H

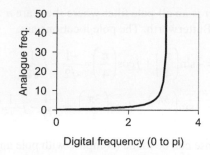

FIGURE 4.10 Function for frequency warping

when Ω varies from 0 to π as when the imaginary part of s (that is to say ω) varies from 0 to infinity. A frequency warped transform function is formed:

$$H(s) = \frac{(F(\Omega) - z_1)(F(\Omega) - z_2)}{(F(\Omega) - p_1)(F(\Omega) - p_2)} \tag{4.34}$$

This does not look very helpful until we note the definition of $F(\Omega)$ and substitute that into the expression:

$$H(s) = \frac{\left(\dfrac{z-1}{z+1} - z_1\right)\left(\dfrac{z-1}{z+1} - z_2\right)}{\left(\dfrac{z-1}{z+1} - p_1\right)\left(\dfrac{z-1}{z+1} - p_2\right)} \tag{4.35}$$

In other words, by warping the frequency variable:

We have created a z domain transfer function that has exactly the same response over the interval $0 < W < p$ as the s domain transfer function gives over the infinite range of possible values of w.

This process is called *bilinear transformation*, because it transforms the pole/zero map of a prototype analogue filter to an equivalent digital form using a bilinear form ($B(z)$). Bilinear transform can be used directly as a design technique for digital IIR filters. It is also used indirectly in the derivation of some important classes of digital filters that we shall now consider.

4.9.1 BUTTERWORTH LOW PASS FILTERS

Of the many types of analogue filters, some standard forms are particularly useful as they preserve their attractive features under bilinear transformation to a discrete time implementation. One such filter type is the Butterworth low pass filter, which has a flat frequency response in the pass-band. The poles in the s-plane for an analogue Butterworth filter are:

$$s_r = -\sin\left(\frac{(2r-1)\pi}{2n}\right) + j\cos\left(\frac{(2r-1)\pi}{2n}\right) \tag{4.36}$$

with $r = 1, 2, \ldots n$

The Butterworth filter is an all-pole filter where there are n zeros at infinity. Consider a 2nd order Butterworth. The pole locations are:

$$s_1 = -\sin\left(\frac{\pi}{4}\right) + j\cos\left(\frac{\pi}{4}\right) = \frac{-1}{\sqrt{2}} + j\frac{1}{\sqrt{2}}$$

$$s_2 = -\sin\left(\frac{3\pi}{4}\right) + j\cos\left(\frac{3\pi}{4}\right) = \frac{-1}{\sqrt{2}} - j\frac{1}{\sqrt{2}}$$

(4.37)

The frequency response can be found (compare with pole and zero representation in z-plane, which is similar):

$$F(s) = \frac{1}{\left(s + 1/\sqrt{2} - j/\sqrt{2}\right)\left(s + 1/\sqrt{2} + j/\sqrt{2}\right)} = \frac{1}{s^2 + \sqrt{2}s + 1}$$

(4.38)

We need to consider a specific −3dB frequency to go further with the example (the above has $\omega_{-3dB} = 1$). Consider this to be ¼ of the sampling frequency. Using Equation 4.33), this makes $\Omega = 2\pi(1/4)/1 = \pi/2$. Remembering that the bilinear transformation is a non-linear process, it is necessary to pre-warp the cut-off frequency of the filter so it appears at the right frequency after the non-linear bilinear transformation is applied. So, effectively, the analogue cut-off frequency ω_a is now given by:

$$\omega_a = \tan(\Omega/2) = \tan(\pi/4) = 1.0$$

(4.39)

At this point, the frequency of the cut-off is applied to the transfer function in Equation 4.38:

$$F(s) = \frac{1}{\left(\dfrac{s}{\omega_a}\right)^2 + \sqrt{2}\left(\dfrac{s}{\omega_a}\right) + 1} = \frac{1}{s^2 + \sqrt{2}s + 1}$$

(4.40)

Now applying Equation 4.31, substitute $s = (z-1)/(z+1)$:

$$H(z) = \frac{1}{\left(\dfrac{z-1}{z+1}\right)^2 + \sqrt{2}\left(\dfrac{z-1}{z+1}\right) + 1} = \frac{(z+1)^2}{\left(2 + \sqrt{2}\right)z^2 + 2 - \sqrt{2}}$$

(4.41)

From this, you can get the difference equation and the pole-zero locations:

$$H(z) = \frac{\dfrac{1 + 2z^{-1} + z^{-2}}{\left(2 + \sqrt{2}\right)}}{1 - \dfrac{2 - \sqrt{2}}{\left(2 + \sqrt{2}\right)}z^{-2}}$$

(4.42)

$$a_1 = 1, a_2 = 0, a_3 = 0.1716, b_1 = 0.2929, b_2 = 0.5858, b_3 = 0.2929$$

These tally exactly with the output from $[b, a] = \text{butter}(2, 0.5)$

If you feel strong, you could go back over the derivation keeping the order of the filter and the cut-off frequency general; you should end up with the following equations. The digital implementation of a Butterworth low pass filter has n zeros at $z = -1$. The n poles p_m lie on an elliptical locus in z, defined by the following equations, in which PR_m and PI_m represent the real and imaginary components of P_m:

$$PR_m = \frac{\left(1 - \tan^2\left(\Omega_1/2\right)\right)}{d} \tag{4.43}$$

$$PI_m = \frac{2\tan\left(\Omega_1/2\right)\sin(m\pi/n)}{d} \tag{4.44}$$

$$m = 0,1,.....(2n-1) \tag{4.45}$$

$$d = 1 - 2\tan\left(\Omega_1/2\right)\cos(m\pi/n) + \tan^2\left(\Omega_1/2\right) \tag{4.46}$$

If n is even, the terms $(m\pi/n)$ are replaced by $(2m+1)\pi/2n$.

Only the poles inside the unit circle are used.

Example

1st order Low Pass filter ($n = 1$), cut-off at $\omega_i T = 0.25\pi$

```
m = 0 ....
d₀ = 0.3431
PR₀ = 2.414  (do not use as outside unit circle)
PI₀ = 0
m = 1 ...
d₁ = 2
PR₁ = 0.414
PI₁ = 0
So: 1st order zero at z = -1, pole at z = 0.414
```

The MATLAB script *Butterworthbilinear.m* does this calculation. The signal processing toolbox in MATLAB also has these types of functions.

4.10 FIR FILTER DESIGN—THE FOURIER TRANSFORM METHOD

Practically, how do we design FIR filters to achieve a desired frequency response? To answer this question, we use the relationship between the impulse response sequence b_j and the response function $H(\omega)$ implied by Equation 4.1. We specify a desired transfer function, $H_d(\omega)$, and work back towards an impulse response. Since the impulse response elements are the coefficients of the difference equation (Equation 4.2), which implements the filter, this represents a practical design strategy. The technique relies upon the inverse transform required to move from a statement of desired frequency response function to impulse response and, so, is called the "Fourier transform method" of designing FIR digital filters.

In summary, the Fourier transform method works as follows:

- Specify your desired frequency response function
- Inverse transform the desired response to give the filter impulse response.

Unfortunately, this is easier said than done. Understanding the Fourier transform design method requires that we consider some important issues.

4.10.1   TIME/FREQUENCY EFFECTS

There is a reciprocal relationship between time and frequency: short in time is broad in frequency (and vice versa). We have discussed this effect before in Chapter 1 and with windowing in this chapter; we called it the inverse spreading theorem. This has a very significant impact on the practical implementation of FIR filters.

Example

The steeper the filter's roll off, the greater the reject rate, the greater the number of coefficients (i.e. the longer in time the impulse response must be).

4.10.2   LEAST SQUARE ESTIMATES OF TRANSFER FUNCTIONS

Practical FIR filters have to be of finite order (i.e. they have a finite length impulse response) and so they cannot accurately match very sharp effects in frequency (such as very fast roll-off rates). This compromise between the order (and, thus, computational cost) of an FIR filter and the type of frequency domain effect it can implement, eventually motivates the use of IIR digital filters.

If, however, we design an FIR filter $H(\omega)$ using the Fourier transform method, we are making a filter that, for the finite order (number of coefficients) chosen, gives the best least square approximation to the target response $H_d(\omega)$, in the sense that error e:

$$e = \sum_{j=1}^{N} \left| H(\omega) - H_d(\omega) \right|^2 \tag{4.47}$$

is minimised.

Note that this is only one definition of a good filter (one in which the response at each frequency is considered equally important), and a least square error design, obtained from the Fourier transform method, may not be the best filter for all applications.

4.10.3   PRACTICAL FILTERS HAVE REAL COEFFICIENTS

The basic difference equation describing an FIR filter (Equation 4.2) is assumed to have real coefficients; indeed, anything other than real coefficients means nothing! This has direct consequences upon the type of desired frequency response we should specify.

An important property of Fourier transforms; the Fourier transform of a real sequence has even real parts and odd imaginary parts, such that:

$$H(\omega) = H^*(-\omega) \tag{4.48}$$

So the negative frequencies we ignored earlier in Fourier analysis come back to haunt us!

4.10.4 Zero Phase and Linear Phase Filters

If we are specifying a target magnitude frequency response that we would like our digital filter to achieve, we could specify a real function of frequency, leaving the imaginary component equal to zero. This generates a filter with zero phase response, called, not surprisingly, a *zero phase filter*. Unfortunately, a zero phase filter cannot (usually) be implemented in real time as it is not causal (it responds to future inputs rather than past inputs).

Fortunately, a zero phase set of filter coefficients can be transformed to a filter that is physically realisable very easily; if the impulse response is delayed, such that the filter is causal, the magnitude frequency response is not influenced. The phase response of such a filter is linear since the delay operation introduces a linear phase term to the frequency response:

$$\angle H(\omega) = e^{-j\omega\Delta} \tag{4.49}$$

where Δ is the time delay.

Such filters are called linear phase filters (and are of particular importance in audio, where the effects of the delay can be compensated for easily).

Run the MATLAB demonstration script *FIR_design1024.m*. This demonstrates linear and zero phase designs. There is only one problem with the design, and that is it takes a lot of time for a signal to pass through (the coefficients cover 1024/11025 = 0.09 seconds), so this is quite useless for real-time processing as the filter would introduce a 0.09s delay.

Run the MATLAB demonstration script *FIR_design32.m*. This uses a more realistic number of filter coefficients (32). Concentrate on MATLAB Figure 5. This compares the desired frequency response (H_d) with that actually achieved (H). We can see our rather idealised "infinite reject band, infinite roll off, no ripple" filter cannot be achieved. In fact, we get significant ripple, particularly noticeable in the reject band.

Why? We need an infinite number of coefficients to represent our idealised brick wall filter exactly. By taking a finite number of coefficients, we are only using a truncated set of the true filter. This should start sounding familiar. It is the same as when we discussed the issue of taking infinite and truncated Fourier transforms of signals. In terms of equations, our "ideal" filter would be:

$$y(t) = x(t) \otimes h(t)$$

$$y_k = \lim_{N \to \infty} \left\{ \sum_{i=0}^{N-1} h_i x_{k-i} \right\} \tag{4.50}$$

But, in reality, we are taking a truncated filter, $h_T(t)$

$$y(t) = x(t) \otimes h_T(t)$$

$$y_k = \sum_{i=0}^{L-1} h_i x_{k-i} \qquad (4.51)$$

(h is used for the coefficients here instead of b).

Now the truncated filter is of length L time bins, which occupies T seconds. This is formed from the infinite filter by a simple rectangular windowing function $w(t)$:

$$\begin{aligned} w(t) &= 0 \qquad t < 0 \; or \; t > T \\ w(t) &= 1 \qquad 0 \le t \le T \end{aligned} \qquad (4.52)$$

and $h_T(t) = h(t) \cdot w(t)$

So combining Equations 4.8 and 4.9, our real filter function becomes:

$$y(t) = x(t) \otimes (h(t) \cdot w(t)) \qquad (4.53)$$

Now we are designing to a specified transfer function, so we should look at this. Putting this into the frequency domain by taking a Fourier transform (remembering multiplications become convolutions and vice versa):

$$Y(\omega) = X(\omega) \cdot H(\omega) \otimes W(\omega) \qquad (4.54)$$

So our actual transfer function is going to differ from our desired transfer function. This can be set out mathematically as:

$$H(\omega) = H_d(\omega) \otimes W(\omega) \qquad (4.55)$$

Now $w(t)$ is a rectangular window, and from our Fourier work you may remember that this means $W(\omega)$ is a sinc function. Run the MATLAB script *sincdemo.m* to confirm this and look at MATLAB Figure 1.

Now $W(\omega)$ has a major lobe (big peak in middle) and a whole series of side lobes (peaks either side of the big peak). It is this that introduces the ripple into the frequency response.

Consequently, a better filter design would be achieved if we used a $w(t)$ that had a "better" Fourier transform. The ideal $W(\omega)$ would be a delta function, because then:

$$H(\omega) = H_d(\omega) \otimes W(\omega) = H_d(\omega) \otimes \delta(\omega_0) = H_d(\omega) \qquad (4.56)$$

i.e. our desire would be an exact match the actual frequency response. But we cannot achieve a pure delta function; the best we can do is to produce a large, narrow major lobe and very low side lobe levels. Look at Figure 2 of MATLAB *sincdemo.m*; here is a time function $w(t)$ that achieves something close to the desired delta function $W(\omega)$. And (surprise, surprise) it is a windowing function (in this case a Hamming window) just like we used with truncated Fourier transforms.

So now that we know what $w(t)$ to use, return to our FIR filter design with 32 coefficients. Run the MATLAB script *a_better_FIR_design32.m*. MATLAB Figure 1 shows the frequency response with a rectangular and Hamming window. Also, the desired response is shown. By using a window, we improve the reject rate by about 20dB, but we trade this against a slower roll off. MATLAB Figure 2 shows the coefficients. A key point to note here is that a truncated shape assumes that outside the graph the coefficients are zero. This is fine for the Hamming windowed coefficients as they go smoothly down to zero. It is a problem, however, for the rectangular windowed coefficients because there is a discontinuity (look carefully at the edges—it does not have a smooth gradient). This sharp edge is what causes problems with the filter.

4.10.5 RECAPITULATION: FIR FILTER DESIGN PROCEDURE

The following procedure will produce n coefficients.

- Form the ideal filter function $H_d(\omega)$ with the desired frequency response with m positive frequency points ($m >= n$). This must include both positive and negative frequencies related by complex conjugates stacked in the normal array format.
- Either
  - Calculate the inverse FFT or DFT to get $h(t)$, the impulse response.
  - Shift the impulse response to ensure when you truncate that all important parts will remain.
  or
  - Multiply $H_d(\omega)$ by $e^{-j\omega\Delta}$ to facilitate time shift by Δ in time domain.
  - Calculate the inverse FFT or DFT to get $h(t)$, the impulse response.
- Truncate (if $m < n$) to take most important coefficients.
- Use windows say a Hamming window, to reduce truncation effects (also known as Gibb's phenomenon).

Of course, there is a MATLAB function to do this design: fir1. If you understand the procedure, you can be an intelligent user of this MATLAB function.

In many applications, the major concern is the reject rate rather than the detail of the frequency response. However, there are applications where filter design is exacting. In those cases, there are methods that allow much more (easy) control over the filter characteristics, and you can find them implemented in the MATLAB signal processing toolbox.

SUMMARY

Filters are synthesised systems to achieve desirable design objectives. Filter design represents the major part of classical digital signal processing. Although there are dozens of digital filter implementations, they can be categorised into two general types, namely FIR and IIR filters. The FIR filters have impulse responses of limited lengths, while the impulse responses of IIR filters are infinitely long. From a perspective of filter structure, FIR filters take a feed-forward structure; IIR filters employ recursive configurations or feed-back loops. This chapter has presented basic concepts of these two types of digital filters through some intuitive examples. Filter design is a vast area; the content included in this chapter is by no means comprehensive. The purpose of this chapter is to outline essential concepts and provide a starting point of filter design. With these concepts, readers can make use of filter design tools, such as those available on the MATLAB platform, to design filters for their application needs. For readers interested in more in-depth understanding and the state-of-the-art of filter design, they are referred to specialist texts.

BIBLIOGRAPHY AND EXTENDED READING

Hayes, M. H. (2008) *Statistical Digital Signal Processing and Modelling*, Wiley.
Mulgrew, B. (2002) *Digital Signal Processing: Concepts and Applications*, Palgrave.
Oppenheim. A. V. and Schafer R. W. (2013) *Discrete-Time Signal Processing: Pearson New International Edition*, Pearson.
Oppenheim, A. V. and Willsky, A. S. (2013) *Signals and Systems: Pearson New International Edition*, Second Edition, Pearson.
Rabiner, L. R. and Gold, B. (1975) *Theory and Application of Digital Signal Processing*, Prentice Hall.
Smith, J. O. (2007) *Introduction to Digital Filters: With Audio Applications*, W3K Publishing.

EXPLORATION

1. Download the file *filters.zip* and unpack it in working directory of MATLAB. This zipped file contains a readme file and heavily commented example codes that include the ones mentioned in this chapter and those for further exploration below.
2. Some suggested further explorations are listed below.

FIR FILTERING

The script *anFIR.m* illustrated a simple filtering process for a low pass filter. The coefficients, b, represent the systems impulse response, so the easiest way to get the filters frequency response is by carrying out an FFT of b. You will get clearer results if you use fft(b,1024) i.e. zero pad the impulse response with a lot of zeros. Put a frequency axis on the graph. You should find a low pass filter with a −3 dB point of about 1.65 kHz if you have everything right.

    a.   Look at the effect of varying the length of the filter (number of coefficients) on the filter characteristics, such as ripple, roll off, reject rate. etc. i.e. vary the value of n taps. Do longer or shorter filters have a better frequency response? What process could be used to improve the performance of the filters (consider Gibbs's phenomenon)?

    b.   For a fixed number of coefficients, vary the number "factor" on line 12. Look at the effect this has on the frequency response and the graph of filter coefficients. What important filter characteristic is this number determining? And what feature of the filter coefficient graph is this related to?

BILINEAR HIGH PASS

Load the script *demo_bilinear.m* and read the notes concerning the bilinear design of a high pass filter. The demo will allow you to move the pole along the real axis and see the effect on the transfer function. A pole-zero plot is also given. As the pole approaches the zero, does the −3 dB point decrease or increase in frequency? Examine the effects also on the impulse response. As the pole approaches the zero, does the impulse response lengthen or shorten, and what would be the significance to audio quality (if any)?

BIQUADRATIC NOTCH

Devise a script that demonstrates the biquadratic IIR design. You can use the same principles as shown in *demo_bilinear.m*. Design a notch filter. You can check your answer against mine, the right graph was generated to have a notch at $f_s/8$ and the poles were at a radius of 0.75 away from the origin. Play with the positions of the poles and zeros and determine the effects.

BIQUAD BAND PASS

Devise a script that demonstrates the biquadratic IIR design. You can use the same principles as shown in *demo_bilinear.m*. Design a band pass filter. Play with the positions of the poles and zeros, and determine the effects.

BILINEAR TRANSFORM

In this chapter, a design method for a Butterworth filter using the bilinear transform was developed. Another common filter is the Chebyshev. The Chebyshev filter has more ripple in the pass band, but the reject (stop) band performance is better. The magnitude function is given by:

$$|H(\omega)| \approx \frac{1}{\left[1+\varepsilon^2 C_n^2\left(\dfrac{\omega}{\omega_1}\right)\right]^{0.5}} \qquad \text{Equation 1}$$

where C_n is the so-called Chebyshev polynomial of nth order. ω_1 is the nominal cut-off frequency. The amount of ripple in the pass band is related to ε. For zero order and first order filters, the Chebyshev polynomials are:

$$C_0(x) = 1$$
$$C_1(x) = x$$

Equation 2

Second and higher-order polynomials may be generated from the following relationship:

$$C_n(x) = 2xC_{n-1}(x) - C_{n-2}(x) \qquad \text{Equation 3}$$

The location of the poles and zeros are given by:

$$PR_m = \frac{2(1 - a\tan(\Omega_1/2)\cos(\varphi))}{d} - 1 \qquad \text{Equation 4}$$

$$PI_m = 2b\tan(\Omega_1/2)\sin(\varphi)/d \qquad \text{Equation 5}$$

$$d = \left[1 - a\tan(\Omega_1/2)\cos(\varphi)\right]^2 + b^2\tan^2(\Omega_1/2)\sin^2(\varphi) \qquad \text{Equation 6}$$

$$a = 0.5\left(c^{1/n} - c^{-1/n}\right) \qquad \text{Equation 7}$$

$$b = 0.5\left(c^{1/n} + c^{-1/n}\right) \qquad \text{Equation 8}$$

$$c = \left(1 + \varepsilon^{-1} + \varepsilon^{-2}\right)^{1/2} \qquad \text{Equation 9}$$

where $\phi = m\pi/n$ and $m = 0,1, \ldots (2n-1)$. If n is even, the terms $(m\pi/n)$ are replaced by $(2m+1)\pi/2n$. The poles of a Chebyshev filter lie on a cardioid shape, and there are n zeros at $z = -1$.

Implement the above equations in MATLAB. Design a Chebyshev filter: 5th order low pass filter ($n = 1$), cut-off at $\omega_1 T = 0.25\ \pi$, $\varepsilon = 0.3298$ (which gives about –3 dB ripple).

Compare your results to a standard MATLAB design. The following lines of code will make and plot the filter.

```
[B A]  = cheby1(5,0.5,0.25,'low')
[H W]  = freqz(num,den)
HH = abs(H(1))
H = 20*log10(abs(H/HH))
hold on
plot(W,H,'r')
[z,p,k] = tf2zp(B,A)
zplane(z,p)
```

Your results will be similar, but not identical.

IIR FILTERING

The script *IIR.m* was a very simple IIR filter. Pass an impulse (Kronnecker delta function) through the filter. This is defined at $x_k = 1$ for $k = 1$, and $x_k = 0$ elsewhere. The output will then be the impulse response of your filter. Take a Fourier transform to look at the frequency response. Is this a low pass or high pass filter? (NB you won't be able to identify a –3 dB point, the filter has such a slight roll off.)

FIR DESIGN

Design a high pass FIR filter with the following characteristics:

- Sampling frequency = 11025 Hz,
- Number of coefficients = 32,
- Cut-off frequency 2000 Hz.

Compare your results to the MATLAB function "fir1."

5 Audio Codecs

5.1 AUDIO CODECS

"Codec" is a combined word of "encoder" and "decoder." An encoder converts or encodes analogue audio signals into a digital format, i.e. codes, and a decoder converts the digital representation of audio back to analogue audio signals. Encoders and decoders are often in pairs, since for a particular encoding process, an associated decoder is needed to convert it back to analogue format; hence, the combined word. An audio codec can be implemented as a silicon chip or a functional block within a chip, or a set software algorithms.

Many audio codecs have been developed; there are still continuous efforts to develop new and even better ones. The performance of these codecs varies in different application scenarios. An analogue or digital converter (ADC) and a digital to analogue converter (DAC) are always needed in any codec, since they link the analogue to the digital signals. The simplest and essential codec is literally the ADC and DAC, and such an encoding method is the well-established pulse-code modulation (PCM), in which the amplitude of audio signals are sampled at uniform intervals and quantized linearly to the nearest available values depending upon the number of bits. PCM encoding and decoding, used in many applications such as the CD encoding format, can be viewed as the "raw" format of digital audio. The sampling rate of PCM encoding determines the frequency range, and the number of bits determines the quantization error, which presents as quantization noise. More sophisticated audio codecs have compression schemes built in. Compression enables the significant reduction of data used to represent the audio signals, and facilitates the storage and transmission of audio information.

The PCM audio codes show redundancy in them. Data compression can be used to reduce the size of the audio file, and the compressed codes may be totally recovered through the decompression. Such an audio compression is called non-lossy compression, since the compression is reversible and causes no information losses or any quality degradation. However, non-lossy compression has limited compression ratios. To further reduce the size of audio files, psychoacoustic models are used in the compression process so that content that will not be perceived or has relatively minor impact on perceived sound quality are removed. Compression schemes following such a process certainly cannot recover the original uncompressed data; therefore, they are lossy compressions.

Because of the differences found in codecs, there are several popular digital audio formats. For each individual format, there are several implementations from different developers. Their performance varies. More sophisticated codecs to allow lower bit-rate and higher perceived quality is an area with much on-going research. Critical evaluation of codecs represents another important branch of the research.

5.2   QUANTIZATION AND PCM FAMILY ENCODING

PCM codecs are relatively straightforward. In PCM encoding, following anti-aliasing filtering and time domain discretization discussed in Chapter 2, a quantization stage is followed.

A finite length section of a general signal, say a voltage $v(t)$, will always be bounded in amplitude by a minimum and maximum value, such that:

$$v_{min} \leq v(t) \leq v_{max} \tag{5.1}$$

for all times t. The voltage then spans a range R:

$$R = v_{max} - v_{min} \tag{5.2}$$

In the process of amplitude quantization, the amplitude range is divided into a number N of equal-width segments, each of width Q:

$$Q = R/N = 2^n \tag{5.3}$$

where Q is the quantization internal and n the number of bits. Each of the N amplitude segments is identified by a unique code, a dimensionless index. With knowledge of the quantization interval Q and some voltage reference point, each code can be related to one of N discrete voltage levels (quantum levels) within the range R. In amplitude quantization, the discrete quantum amplitude level closest to the instantaneous value of the signal to be quantized is used to represent the signal's amplitude at that instant.

Because the amplitude range R is subdivided into elemental divisions of finite width Q, it is obvious that the representation of a continuous variable by the nearest quantization level potentially introduces an error; this error is called quantization distortion or error. Effectively, it is a rounding error and is distortion because it depends on the input signal.

5.2.1   QUANTIZATION AS A NOISE SOURCE

Whatever type of quantizer is used, we have noted that quantization introduces errors. The difference between the true value of a signal and the quantized approximation of the signal represents an unwanted "noise" process. Intuitively, the accuracy of a quantizer will increase as the number of quantizing levels, N, increases as each voltage can be more accurately represented.

We assume that N is "reasonably large" and the input signal $v(t)$ is reasonably uniformly distributed over its range R. These are not unreasonable assumptions for an appropriately designed quantizing system. In an ideal quantizer, the difference between the nearest quantizing level and the instantaneous value of $v(t)$ is never greater than $\pm Q/2$. To calculate statistics of the quantizing noise, we can use the probability density function (PDF).

If $v(t)$ is a random process, it is reasonable to assume that the quantizing noise is independent of $v(t)$ and that it is uniformly distributed over the range $-Q/2$ to $Q/2$. In this case, the quantizing noise's PDF, $p(v)$, is given by:

$$p(v) = \begin{cases} 1/Q, & -Q/2 < v < \dfrac{Q}{2}v \\ 0, & elsewhere \end{cases} \tag{5.4}$$

The mean square value of the quantizing noise is obtained from the second moment of its PDF:

$$MS(v_{quant}) = \int_{-\infty}^{+\infty} x^2 p(x)\,dx = \frac{1}{Q} \int_{-Q/2}^{+Q/2} x^2\,dx = \frac{Q^2}{12} \tag{5.5}$$

Many practical implementations of quantizing systems, however, identify the highest quantizing voltage that is less than the instantaneous signal. This is equivalent to always rounding down rather than rounding to the nearest quantization level. In these systems, the quantizing noise is bounded between $-Q$ and 0, and the quantization error increases by 6 dB (in comparison to the standard "textbook" value for ideal quantization noise).

5.2.2 Quantization as a Distortion Process

It is important that quantization errors are small, but problems arise because this is distortion, not noise. The distortion added to the recording depends on the signal. For example, a square wave whose peak-to-peak amplitude coincides with the output digital level gives an encoding error of zero; if the peak-to-peak amplitude was in the middle of two digital levels, the magnitude of the error would always be a maximum $\frac{1}{2}Q$. In other words, the error is deterministic, not a characteristic expected of a random noise signal.

As with any distortion, the effect is to add harmonics to the original signal (which are not removed by anti-aliasing filters as the distortion occurs after that filter). The high frequency harmonics above the Nyquist limit fold back to stay within the digital systems frequency range ($0\,Hz - f_s/2$), The folded-back harmonics are known as anharmonics and are shown in Figure 5.1. It shows that the second and third harmonics are folded back to stay within the Nyquist limit (aliasing). These are now anharmonics as they are no longer in a simple harmonic series.

Furthermore, for very low-level noise that is hovering around the lowest few quantization ranges, the natural random fluctuations cause the noise to effectively switch on and off in the digital domain. So the noise can be heard switching in and out. This makes a pure quantizer unsuitable for audio applications. Instead, a system of dither is used.

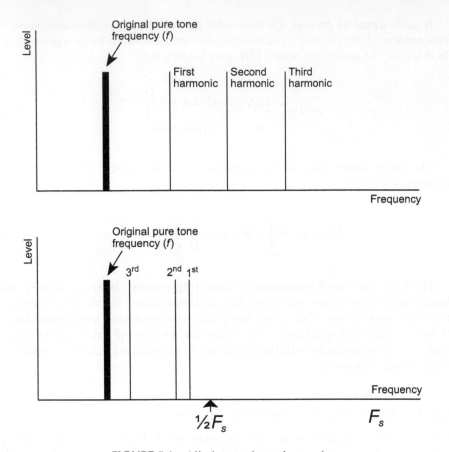

FIGURE 5.1   Aliasing causing *anharmonics*

5.2.3 Dynamic Range due to Quantization

Now that we have expressions for the mean square value of the quantizing noise in practical quantizers, it is possible to calculate the dynamic range of these devices.

Definition: The dynamic range of a system is the ratio of the mean square level of the largest signal to the mean square level of the noise present in the absence of input.

Case 1—Full Scale Sinusoidal Input

If the input to the quantizing process is a sinusoid that covers the entire range of the quantizer:

$$v(t) = \frac{R}{2}\sin(\omega t) \tag{5.6}$$

This has a mean square value of $R^2/8$.

The dynamic range of the quantizer is, therefore:

$$10\log_{10}\left(\frac{R^2/8}{Q^2/12}\right) = 20\log_{10}\left(N\sqrt{6}/2\right) = 20\log_{10}\left(2^{n-1}\sqrt{6}\right) \tag{5.7}$$

n	Dynamic Range
8	49.9 dB
12	74.0 dB
16	98.1 dB

It is seen from this simple analysis that converters having at least 16 bit output are required for serious audio application. The 12 bit converter is used in communication systems and some instrumentation applications and the 8 bit converter is useful only in restricted low-resolution applications.

Case 2—Full Scale Input of Crest Factor c

The sinusoid used as the input signal in the example above is not very representative of a general signal. Audio signals, for example, have transient bursts that give them a very different PDF than that of a sinusoid—they have a higher crest factor.

Definition: The crest factor is the ratio of the signal's peak value to its root-mean-square (RMS) value.

If a quantizer is designed to accept signals $v(t)$ of crest factor c, then the quantizing range R must be chosen so that the signal peaks at $R/2$. If the range is any smaller, the quantizer will clip the signal.

If the peak value of the input is $R/2$, then, by definition, the RMS value must be $R/2c$, such that the mean square value of the signal for an n bit quantizer is:

$$RMS(v(t)) = \frac{R^2}{4c^2} = \frac{Q^2 2^{2n}}{4c^2} \tag{5.8}$$

which can be formed into a dynamic range of:

$$10\log_{10}\left(\frac{Q^2 2^{2n}/4c^2}{Q^2/12}\right) = 10\log_{10}\left(\frac{2^{2n}3}{c^2}\right) = 20\log_{10}\left(\sqrt{3}2^n/c\right) \tag{5.9}$$

This is evaluated for different converter resolutions and input crest factors below. (Note: A sinusoid has $c = 1.41$.)

n	Dynamic Range (dB)		
	$c = 1.41$	$c = 4$	$c = 16$
8	49.9	40.9	28.8
12	74.0	65.0	52.9
16	98.1	89.0	77.0

A music or general communication signal is seen to significantly degrade the dynamic range of the quantizer.

5.3 DITHER

As discussed above, the quantization distortion is often related to the signal and shows certain patterns. Dither is used to remove the audible artefacts of quantization distortion. Dither is a system of using noise to make the process of quantization less deterministic. For example, noise can be added deliberately to a signal before quantization. Although this reduces the signal-to-noise ratio of the system slightly, it removes the problems of the noise appearing at discrete harmonics and anharmonics by smearing the quantization error to appear as broadband noise. We will look at this non-subtractive dither.

Dither noise fluctuates by at least a quantization interval ($\pm\frac{1}{2}Q$), which means that the quantization error is more random (due to the noise), and not deterministic (signal-dependant distortion).

Consider the two slowly varying signals shown in Figure 5.2. The bottom signal produces little quantization error as it lies near a digital level (digital levels marked as horizontal lines). The top signal causes maximum errors as it lies half a quantum from a digital level (we assume that the quantizer rounds to the nearest level). Ideally we would like this signal to produce the same error regardless of where it is with respect to the quantization ranges, i.e. error is signal independent and so is noise, not distortion. This is achieved by adding a small amount of noise to the signal (Figure 5.2). Now the average error in both cases is the same—it is (mostly) signal independent. (It is not completely signal dependent because the variance of the error is different in both cases—see the bibliography at the end of this chapter.)

The addition of dither noise reduces the signal-to-noise ratio. Dither is about compromises. Distortion is much more of a problem than noise in an audio system. Consequently, even though the amount of noise in the system goes up with dither, we can reduce the audibility as it is no longer signal-dependent noise. The reduction in the *S/N* ratio depends on the type of dither used; typically, reductions are in the order of 3 to 6 dB.

Gaussian white noise is probably the simplest noise to use for dithering the signals; however, it is not necessarily the most effective one. Lipshitz and Vanderkooy

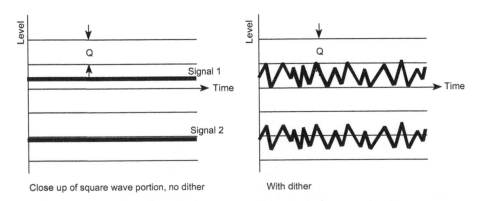

Close up of square wave portion, no dither   With dither

FIGURE 5.2   The use of dither

pointed out that different noise types with different probability density functions (PDFs) behave differently when used as dither signals, and they suggested optimal levels of dither signal for audio (Lipshitz et al., 1991; Vanderkooy and Lipshitz, 1987). Essentially, Gaussian noise needs to be applied at a relatively higher level for full elimination of perceivable distortion than the rectangular PDF or triangular PDF noises. Triangular PDF noise also reduced the so-called noise modulation, i.e. perceivable noise level change in a quiet music passage or silence.

It is interesting to note that with sufficient bit depth, added dither sometimes becomes unnecessary since the noises generated by the microphone and pre-amplifiers might be sufficient to dither the quantization distortion automatically. In addition to white noise, coloured dither is also used on occasion. The so-called coloured dither is a dither signal filtered to have no flat spectrum. The purpose of using the coloured dither is to lower the dither energy in the critical audio bands so that the dither can be made non-audible even though a slightly higher level of dither is applied.

5.4   FROM PCM TO DPCM

So far we have discussed sampling a system in pulse-code modulation (PCM), where the quantization code is proportional to the instantaneous input voltage. Differential pulse-code modulation (DPCM) is formed from the difference between adjacent PCM signal values. If the PCM signal is PCM $x_1, x_2, x_3, \ldots x_n$, then the DPCM signal is $(x_2 - x_1), (x_3 - x_2) \ldots (x_n - x_{n-1})$. DPCM is effectively numerical differentiation of a PCM signal, hence the name. DPCM is fundamental to noise shaping used in oversampling.

As we are now converting the difference between adjacent samples, rather than the samples themselves, there is no longer a limit to how large a signal the system can represent. The limiting factor is, however, how big a difference between adjacent samples it can represent, so a system is limited by its slew rate. To encode large differences between adjacent samples, the convertor can either increase the number of bits that the quantizer produces, or increase the sampling frequency.

If the sample rate is not fast enough, then the system can suffer from *slope overload*, which is when the system is incapable of tracking a fast-changing signal. Figure 5.3 gives an illustration.

The signal output from a DPCM falls at 6 dB/octave compared to a PCM signal. The noise floor due to quantization error and dither is constant across all frequencies.

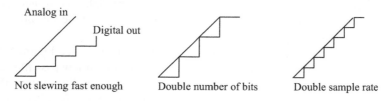

FIGURE 5.3   Illustration of tracking signal changes

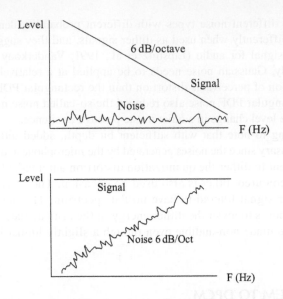

FIGURE 5.4   Signal rolling off (top) and noise floor change after compensation (bottom)

Obviously, this signal roll off (Figure 5.4, top) is undesirable, so the signal is integrated (Figure 5.4, bottom). This tilts the noise floor, so noise is greater at high frequencies; hence, the sigma DPCM converters are known as noise-shaping converters. If this shifts the noise above the audible frequency range (>20 kHz), then a low pass filter can be used to remove this noise. This noise reduction is a primary advantage of an oversampling system.

5.5   OVERSAMPLING AND LOW BIT CONVERTERS

Figure 5.5 illustrates that, in a digital system, we can play off the number of bits against the sampling frequency. If we lower the number of bits representing a signal, we can still get an accurate rendition of the input signal provided we increase the sampling frequency. This fact is the key to the use of oversampling converters, which are commonly used on quality audio systems. Consider a 4x oversampling system for an audio system designed to work to 20 kHz.

Oversampling is used because:

- Analogue anti-aliasing filters are expensive due to the high-quality analogue components needed. In an oversampling system, the analogue anti-aliasing filter can have a very low roll off and be very cheap.
- When oversampled, the noise floor rises with frequency (see Figure 5.4). Consequently, this noise can be filtered out by the digital low pass filter as it is outside the audio frequency range. In this 4x oversampling system, the signal-to-noise ratio is improved by 12 dB. This is a process called noise shaping.

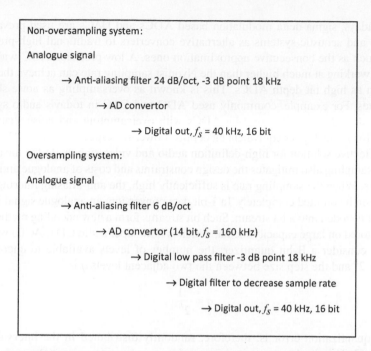

FIGURE 5.5 Comparison of non-oversampling and oversampling systems

5.6 ONE-BIT CONVERSION, SIGMA-DELTA MODULATION

We have discussed through an example how oversampling can help reduce the bit depth of an ADC while maintaining the equivalent bits in terms accuracy. There is no reason why we cannot go to the extreme of only having a one-bit quantizer. This is *delta modulation*. The output is simply +1 or 0 depending on whether the previous signal values are less than or greater than the current value. Thus, the conversion circuit can be very simple. If we then integrate the output from the delta modulator, we get PCM. Such a convertor is known as a sigma-delta modulator, shown in Figure 5.6. So when considering the effects of sampling in an audio system, the simple considerations of quantization error and the simple formulations for dynamic range derived from the PDF become more complicated. The simple formulation assumes that any quantization error produced stays within the audio band and so must be counted, but in oversampling systems, the error is deliberately shifted out of the audio range so that it can be removed by low-pass filtering in the digital domain.

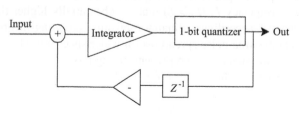

FIGURE 5.6 A simplest form of sigma-delta modulation

Nowadays, sigma-delta modulation based ADCs and DACs are used prevalently in audio and acoustic systems as alternative converters to traditional high-precision ADCs such as the consecutive approximation ones. A low-precision ADC with very few bits working at much higher than the Nyquist sampling rate can achieve the same precision as high bit depth ADCs. This is known as oversampling as noise shaping techniques. For example, commonly used ADC and DAC in today's audio systems and sound cards typically use 1-bit ADCs with oversampling and noise shaping to achieve 16 or even higher equivalent bit depths. Lower-bit, especially 1-bit, ADC offers a cost-effective solution for high-definition audio and video signals. In the meantime, the oversampling also mitigates the design constraints and costs of analogue anti-aliasing filters. When the sampling rate is sufficiently high, the anti-aliasing filtering stage might even be omitted completely. In 1-bit data conversion, an analogue signal is converted or encoded into a bit stream. Such bit streams form a new encoding method and can be stored on large capacity media, which is how super audio CD (SACD) works.

Now consider a B-bit quantizer: the number of levels available to encode the signal is 2^B and the step size between the two adjacent levels q is

$$q = \frac{1}{2^{B-1}} \tag{5.10}$$

The quantization error is, therefore, randomly distributed in the interval from $-q/2$ to $q/2$. This random quantity is viewed as quantization noise $e(n)$,

$$x(n) = x(nT) + e(n) \tag{5.11}$$

where n is sample point and T is sampling interval. The variance of error $e(n)$ becomes

$$\sigma_e^2 = E\left[e^2(n)\right] = \frac{1}{q}\int_{-\frac{q}{2}}^{\frac{q}{2}} e^2(n)\,de = \frac{q^2}{12} = \frac{2^{-2B}}{3} \tag{5.12}$$

Note that the variance of error σ_e^2 is the noise energy, and as the noise is evenly distributed throughout the frequency range, the noise power spectral density is

$$\frac{\sigma_e^2}{f_s} = \frac{q^2}{12 f_s} \tag{5.13}$$

where f_s is sampling frequency. This clearly shows that by increasing sampling frequency f_s, noise power spectrum density can be reduced.

Sampling at a frequency F_s ($F_s > f_s$) that is substantially higher than necessary sampling frequency f_s according to Shannon-Nyquist sampling theorem, we ended up having the quantization noise spread over a wider spectrum while we only need to take a portion of the spectrum to represent the signals.

So when oversampling, the overall noise level is

$$N = \frac{q^2}{12 F_s} \tag{5.14}$$

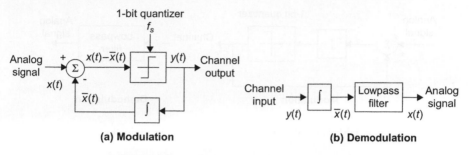

(a) Modulation   **(b) Demodulation**

FIGURE 5.7   Delta modulation and demodulation

and the in-band noise is

$$N_B = \frac{q^2}{12F_s} \cdot \frac{f_s}{F_s} \tag{5.15}$$

This shows clearly how oversampling can reduce quantization noise.

Sigma-delta modulation was originally developed in the 1960s to encode video signals. It was evolved from the delta modulation as shown in Figure 5.7, parts (a) and (b).

Analogue signal is fed into a 1-bit quantizer controlled by a clock f_s. The 1-bit quantizer is simply a threshold and can be implemented by a comparator. The output is taken, integrated, and sent as negative feedback into the input to form a feedback loop. The negative feedback attempts to minimise the errors between input and output; in other words, to force the output to follow the input. The output is a clocked bi-level signal, i.e. a bit stream. The integrator accumulates the errors and makes the overall output equal to input, i.e. attempts to make the statistics of output equal to input. Clearly, an integrator and a low pass filter will recover the original analogue signal. The modulation is based on the error signals, i.e. the input and predicted output; hence, the difference, or delta modulation. The delta modulation is the basis of sigma-delta modulation. As we shall see, an addition, sum, or integration stage will turn the delta modulation into a sigma (sum) and delta modulation. In both modulation and demodulation, an integrator is needed, as illustrated in Figure 5.7. One can move the second integrator to the modulation stage and simplify the de-modulation (Figure 5.8).

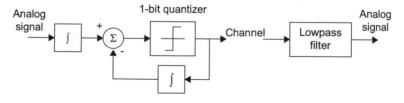

FIGURE 5.8   Sigma-delta modulation (integration before a delta modulator)

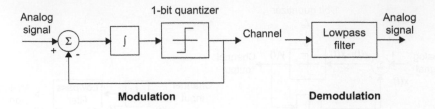

FIGURE 5.9 Sigma-delta modulation and demodulation

By further combining the two integrators, the "standard" sigma-delta modulator and demodulators are shown in Figure 5.9.

Figure 5.10 shows a first-order continuous time model for sigma-delta modulation using Laplace transform, in which independent variable s is the Laplace operator. N is quantization noise.

If $N(s) = 0$, we obtain the output-input relation

$$Y(s) = [X(s) - Y(s)]\frac{1}{s} \tag{5.16}$$

Rearrange to obtain signal transfer function $H(s)$:

$$H(s) = \frac{Y(s)}{X(s)} = \frac{1}{s+1} \tag{5.17}$$

This is a low pass filter.

Let $X(s) = 0$,

$$Y(s) = -Y(s)\frac{1}{s} + N(s) \tag{5.18}$$

The noise transfer function becomes:

$$H'(s) = \frac{Y(s)}{N(s)} = \frac{s}{s+1} \tag{5.19}$$

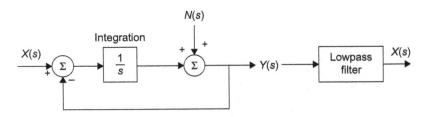

FIGURE 5.10 Sigma-delta model using Laplace transform

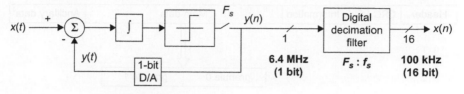

First order ΣΔ loop

FIGURE 5.11   SDM DAC

It is a high pass filter. This sigma-delta modulation suppresses in-band quantization noise and shifts it to higher frequencies. The high-frequency quantization noise can be removed by filtering. This is the noise shaping principle. In more sophisticated sigma-delta modulations (SDMs), the integrator $1/s$ is replaced by a carefully designed filter such that the in-band frequency response is flat, noise level is low. An even more sophisticated filter can incorporate the psychoacoustic model so that the noise is shifted to less sensitive frequency regions of human hearing. Figure 5.11 shows a block diagram of a 1-bit SDM DAC.

Following a sigma-delta modulation loop is a decimation stage. The decimation reduces the sample rate by combining the successive 1-bit data. Noise shaping is a filtering process that shapes the spectral energy of quantization error, typically to either de-emphasise frequencies to which the ear is most sensitive, or separate the signal and noise bands completely. If dither is used, its final spectrum depends on whether it is added inside or outside the feedback loop of the noise shaper: if inside, the dither is treated as part of the error signal and shaped along with actual quantization error; if outside, the dither is treated as part of the original signal and linearises quantization without being shaped itself. In this case, the final noise floor is the sum of the flat dither spectrum and the shaped quantization noise. While real-world noise shaping usually includes in-loop dithering, it is also possible to use it without adding dither at all, in which case the usual harmonic-distortion effects still appear at low signal levels.

5.7   LOSSY CODECS AND MPEG CODECS

Lossy audio codecs are increasingly used in everyday life. All audio heard from digital TV broadcast is compressed in a lossy manner. Lossy audio compression can also be found in digital voice communications, streamed audio, and mobile personal audio devices and other applications where smaller file sizes of lower bit rates are desirable. Lossy codecs employ lossy compression algorithms based on psychoacoustic characteristics of human hearing to mitigate the perceived degradation of audio quality when attempting to reduce the data files and lower the bit rates. Since the encoding is lossy, there is no tangible way to truly recover the

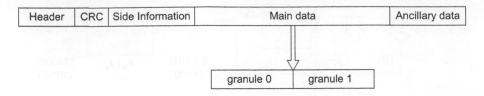

FIGURE 5.12   MP3 frame structure

original files or signals. There are many lossy audio schemes and sub-schemes and associated codecs in popular use. One version of the MPEG is discussed in this section as an example.

The Moving Picture Experts Group (MPEG) is a working group formed by ISO and IEC to set standards for audio and video compression and transmission. MPEG-1 is one of the standards for lossy compression of video and audio by the MPEG. It was originally designed to compress VHS quality digital video and CD audio down to around 1.5 Mbit/s without excessive quality loss for cable transmission and digital broadcasting. MPEG-1 consists of three audio codecs, namely MPEG-1 audio layers I, II, and III. These are defined and standardised in the ISO/IEC 11172-3 (ISO, 1993).

MPEG-1 audio layer I (MP1) is generally obsolete, and replaced by MPEG-I layers II (MP2) and III (MP3). MP2 is currently a main-stream format for digital audio broadcasting due to its relatively simple decoding requirements. MP3 is a popular format for personal audio. The MP3 was instigated by Thomson Multimedia and the Fraunhofer Institute, and then got adopted by the MPEG. Like many other MPEG standards, not every single aspect of the MP3 codecs is specified. Different implementations are possible. Therefore, there are several different but generally compatible implementations.

MP3 encoded audio is a bit stream. It is divided into many frames. Each frame contains 1152 time domain signal samples. A frame is further divided into two equal "granules," each containing 576 samples. A frame has a 32-bit header, which contains synchronisation word and description of the frame. The 16-bit cyclic redundancy check (CRC) code provides an error checking mechanism. Side information can be 16/32 bits for mono and stereo. It carries information about how to decode the main audio data stream. Ancillary data is optional for MP3 and the length is not specified. For a complete MP3 file, a 128-byte metadata tag (ID3) is used to carry text information about title, artists, album, year, track, genre, etc. Figure 5.12 shows the frame structure. The header is followed by the psychoacoustically compressed audio data. Length of audio data can vary.

The encoding used in MP3 has two major compression stages. The first stage is a lossy compression one; it applies a psychoacoustic model to remove information from the signals that is not perceivable or perceptually less important. Audio is filtered into 32 sub-bands and a modified discrete cosine transform (MDCT) is performed to allow for checking the signal with the psychoacoustic model. The second stage is a non-lossy one. It is used to further compress the data using Huffman coding. Figure 5.13 shows a block diagram of an MP3 encoder.

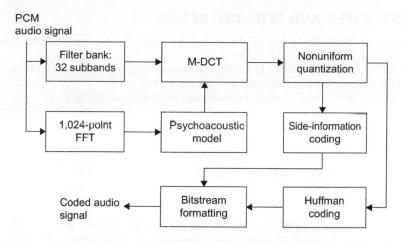

FIGURE 5.13   A block diagram of MP3 encoding

SUMMARY

Audio is about reproduced sound. Digital audio requires that analogue and acoustic signals that represent sound are encoded in a digital format first before any possible subsequent stages such as processing, storage, or transmission. Audio signals in digital formats need to be reconstructed or decoded back to analogue signals and eventually reproduce audible sound. Encoders and decoders are the devices for these tasks, and "codec" is a compound word for encoder and decoder. The simplest codecs are linear and "multi-bit" analogue-to-digital converters and digital-to-analogue converters. The associated digital audio format of these codecs is the so-called PCM codes. There are many other ways audio can be encoded and decoded. Bit stream resulted from a kind of pulse-density-modulation or 1-bit conversion is an example. Although the bit stream is not an efficient way of encoding signals, it is directly related to the 1-bit DACs with sigma-delta modulation. Lossy compression is increasingly used in audio. Employing psychoacoustic models, unperceivable or less-sensitive components in audio signals can be discarded in the encoding to drastically reduce the size of audio files without significant compromise of perceived audio quality; this leads to Lossy encoding and, of course, the associated decoding methods. This chapter discussed the principles and purposes of codecs used in audio, and outlined some basic concepts. Audio codecs, especially compression based on psychoacoustic models, represent a vast and on-going area of research.

REFERENCES

ISO/IEC 11172-3 (1993) Information Technology—Coding of Moving Pictures and Associated Audio for Digital Storage Media at up to about 1,5 Mbit/s -Part 3:Audio.

Lipshitz, S. P., Vanderkooy, J., and Wannamaker, R. A. (1991). "Minimally Audible Noise Shaping," *Journal of the Audio Engineering Society*, Vol. 39, No. 11, pp. 836–852.

Vanderkooy, J. and Lipshitz, S. P. (1987). "Dither in Digital Audio," *Journal of the Audio Engineering Society*, Vol. 35, No. 12, pp. 966–975.

BIBLIOGRAPHY AND EXTENDED READING

Kamenov, A. (2014) *Digital Signal Processing for Audio Applications*, Second Edition, CreateSpace Independent Publishing.
Pavan, S., Schreier, R. and Temes, G. C. (2017) *Understanding Delta-Sigma Data Converters* (IEEE Press Series on Microelectronic Systems), Wiley.
Pelgrom, M. (2018) *Analog-to-Digital Conversion*, Third Edition, Springer.
Pohlmann, K. (2010) *Principles of Digital Audio*, Sixth Edition (Digital Video/Audio), McGraw-Hill/Tab Electronics.

EXPLORATION AND MINI PROJECT

EXPLORATION

1. Downloadable zipped file *codecs.zip* contains a readme file, audio samples, and some MATLAB® codes for demonstration, illustration, and further exploration of the concepts outlined in this chapter. Download and unpack it in the working directory of MATLAB.
2. Perceptual appreciation: There are also two MP3 files for demonstration, *quant_sine_1.mp3* and *quant_sine_2.mp3*. These are sine waves (pure tones) of two different frequencies; each recording is arranged in the sequence of original signal (quantized with a 16 bit ADC), quantized signal (quantized with a 5-bit ADC), and quantization error. Listen to these carefully. Is the quantization noise signal dependent? How would you describe the noise?
3. You have already used the MATLAB script *basescript.m* to create 400 Hz and 10000 Hz (line 20 defines the frequency) sine waves and investigate their spectra, etc. Now we modify this script further to investigate quantization effects.
   - Quantize the signal and observe the effect on the spectrum. It is suggested this can be done by using the round() function (see line 31). Round the sine wave to 1 decimal place to form 21 quantum levels, and a clearly audible and visible effect.
   - Try altering the input frequency a little and observe the change in the harmonic structure. Make sure you test 408 Hz. What is an anharmonic? Use the sound() function to listen to the quantized signal; what does it sound like in comparison to the original?
   - The signal-to-noise ratio of a quantizer was derived in the notes. Calculate the signal-to-noise ratio in MATLAB using the time signal and the error introduced into the time signal on rounding, and compare to expectation. The s/n ratio can be found by getting the mean square signal and mean square error, and taking $10\log_{10}()$ of the ratio. Given a time signal in array x, the mean square can be found in MATLAB using mean(x.*x) (line 32). Note the use of .* (dot times) so the calculation is done array element by array element; a simple * will make MATLAB try and find the square of the matrix).

- Instead of using a sine wave, use a real audio signal. For example, the script *music.m* loads in an audio signal from a WAV format file. Find the signal-to-noise ratio of the quantisation process using this audio signal and compare to the result you found for the sine wave signal. In the notes, a formulation was given for the dynamic range achieved for a given signal crest factor. Test whether this formula works.

4. The following MATLAB code (one line) will generate rectangular probability dither, assuming that the quantisation intervals are 0.1 and that you are rounding to the nearest quantisation interval:

$$d = (\text{rand}(N,1)-0.5)/10.0$$

- Add the dither to a sine wave before quantization and observe the visual effects on the spectra and also the audible effects via a sound card. What is the change in the signal-to-noise ratio? Produce a formulation to predict the signal-to-noise ratio. Assume that the errors introduced by the dither and the quantization process are independent of each other. Listen to the quantized and dither signal; what does it sound like in comparison to the original? Concentrate on the timbre of the sine wave and the noise levels.
- Use a swept sine wave as an input to the quantizer. (MATLAB has a function chirp to generate this). Listen to the quantized signal, the quantization error, and examine the effects of dither. A swept sine signal demonstrates the problems of quantization error quite dramatically.

5. The MATLAB script *oversampling.m* is a basic implementation of an oversampling sigma-delta modulator as shown below:

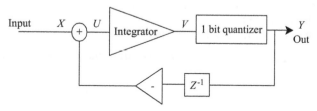

Run and examine the outcomes of the script. There are many things to note, but of primary interest:

- Note that a 1 bit signal does really convey frequency information (Figures 2 and 3 in MATLAB window). This can seem fantastical, but you can see it in the time picture. (If you stare at this long enough, you can also see a picture of the Matterhorn.)
- Figure 3 shows the input frequency and all the additional frequencies generated by quantisation. Notice the rising noise floor due to this being a noise shaping convertor. The frequencies of interest are to the far left of the graph, and so the noise >16 kHz is going to be lost when we down-sample it.

- Figure 4 shows the input and output signals. Notice they are not identical. The sine wave is distorted by the quantisation—we need to add dither.
- Figure 5 shows the final spectrum of the converted signal—the distortion is too large to be useful.
- The simulation can be extended to more complex systems.
- How could dither be fitted into this convertor?

MINI PROJECT IDEA

Audio signals can be pulse-width modulated (PWM) or pulse-density modulated (PDM) to directly drive a switched-mode amplifier, sometimes called commercially a "digital amplifier." The pseudo-code below converts a PCM-coded signal into a PDM signal using a first order SDM. Use it as a blueprint to write some MATLAB codes to encode an audio excerpt into PDM, low pass filter bit stream as appropriate, and play it back through a loudspeaker.

```
// Encode samples into pulse-density modulation
// using a first-order sigma-delta modulator
function pdm(real[0..s] x)
  var int[0..s] y
  var real[-1..s] qe

  qe[-1] := 0                 // initial running error is zero

  for n from 0 to s
      if x[n] >= qe[n-1]
         y[n] := 1
      else
         y[n] := -1
      qe[n] := y[n] - x[n] + qe[n-1]

  return y, qe                // return output and running error
```

6 DSP in Binaural Hearing and Microphone Arrays

Human hearing is naturally binaural since two ears are involved in sensing sounds around a normal listener. Stereophony, in the context of audio engineering, represents the simplest form of technologies that enable source localisation. A typical 2-channel stereo system represents a minimum setup for stereophonic sound stages, and it is relatively straightforward to understand. Surround sound offers more channels for sound engineers to create virtual effects and can arguably mitigate the constraints of listening environments, but it remains essentially a two-dimensional (2D) audio-rendering technique. No tangible, accurate, or complete information about elevation is included. Nonetheless, human hearing can do much more than just 2D sound sensing. Many spatial cues allow for localistion and separation of sources, not only in azimuth but elevation aspects as well. This means that binaural hearing empowered by two ears is actually three-dimensional (3D) hearing capable of sensing the location of sources in the whole space around a listener. Information about source locations is embedded in the two signals that reach the eardrums in terms of their strength, time, and frequency discrepancies. Brains work on these feeble variations and resolve the 3D locations of sound sources effectively.

Human auditory is a complex system. The distance between the two ears, shapes of the head and face, sizes and shapes of ear canals, and the complicated structures of outer ears all contribute to the discrepancies in stimuli upon hair cells inside the ears that convert the acoustic signals to neural stimuli to the brain. The processing power of the brain should never be underestimated. Received signals are further processed by the brain to give 3D spatial sound perception. To date, a powerful digital computer with advanced DSP and machine learning algorithms can only, to a very limited extent, partly emulate such complicated processes. Spatial audio, on some occasions termed virtual audio display (VAD), intends to capture, reproduce, and even synthesise the 3D spatial acoustic cues for many applications. There are also "inverse" problems to be solved, for example, de-composition of spatial sound stage into individual sources and identification of their locations. These are particularly useful for the production of object-based audio and automated sound source monitoring in environmental acoustics.

In this chapter, we discuss signal processing for spatial sound and microphone arrays. This is a vast area of study, since signal processing is no longer restricted to one-dimensional time series analysis and manipulation. The modeling of sound filed in the space, and the interactions of sound waves and obstacles are encountered. The systems under investigation are often multiple-input-multiple-output (MIMO) instead of single-input-single-output (SISO) as we often consider in classical signal processing. In a single chapter, it is impossible to exhaust all aspects of spatial audio and array signal processing, which should justify a whole different specialist text.

This chapter highlights the head related transfer functions (HRTFs) and binaural techniques, ambisonics, and microphone arrays. The HRTFs explain, to some extent (since brain functions are not included), how 3D sounds are perceived by two ears, and binaural hearing represents the simplest way to retain all spatial information that human listeners can acquire through their biological auditory systems. Ambisonics, on the other hand, represents one elegant way to encode sound field information. Microphone arrays and beam-forming techniques offer flexibility and almost unlimited capability to capture sound fields and selectively detect the sources, provided that large and complicated arrays and relevant signal processing are practical. The immediate applications can be the capture and reproduction of audible spatial sounds, but the application can go far beyond this. Advanced microphone arrays can enable much higher resolution and source localisation capability than human ears. Such measurement or acoustic sensors can be used to determine the location of a sound source and focus on interested sound, or enable acoustic echo-based scanning.

6.1   HEAD RELATED TRANSFER FUNCTION AND BINAURAL SIGNAL PROCESSING

High performance audio and video codecs, ever increasing bandwidths of data communications networks, and affordable massive data storages have all contributed to the multi-platform delivery of higher definition media content, and ubiquitous and personalised multimedia content consumption. For a realistic and comfortable user experience, attention should be paid to the acoustics so that the sound and visual images are kept in good agreement. Video conferencing, gaming and other virtual reality applications all call for high definition virtual auditory space (VAS); as a result, 3D spatial audio is sought after. One possible way to implement the VAS is the binaural techniques that render the virtual sound spaces through headphones via digital processing of source signals with HRTFs (Shinn-Cunningham, 1998). On the other hand, if straightforward binaural recording is made and if we are prepared to discount the discrepancies of the HRTFs in individuals, there is no need for processing at all. Indeed, many other spatial audio signal acquisition and especially reddening techniques are related ultimately to human binaural hearing and, hence, the HRTFs. The HRTFs partly explain why and how two ears can achieve spatial hearing, i.e. the peripheral aspect of binaural hearing, which is responsible for the common way normal hearing people listen to various sounds naturally in real life. This section will be a general introduction to the HRTFs and associated simple signal processing techniques. Xie (2013) offers more detailed analysis and in-depth insights into this subject.

As we shall discuss in this section later, HRTFs are related not only to the head, but the outer ears, ear canals, and torso as well. Head and torso simulators (HATS) with calibrated artificial ears are often used to emulate a human subject. A slightly simplified model is the dummy head without torso. A further simplified model can be just a sphere without even the outer ears to emulate the head. The use of dummies to do the measurements is a "one size fits all" approach. The results are twofold: A representative dummy simplifies the measurement procedures, eases the logistical or physical constraints, and mitigates variations (individuality) of measurements. However, research so far has indicated the importance of individual HRTFs. Best results can only be achieved from individualised HRTFs.

6.1.1 HEAD RELATED TRANSFER FUNCTIONS (HRTFs)

It is well established in psychoacoustics and spatial audio that sound source localisation capability of audio perception is due to three key factors: (1) the inter-aural time difference (ITD), which gives directional localisation cues for relatively lower frequencies (<1.5 kHz); (2) the inter-aural level difference (ILD), responsible for relatively higher frequencies (>1.5 kHz); and (3) various spectral cues due to interactions between sound waves and the listener's head, torso, and, in particular, outer ears (pinnae). Dynamic cues due to head movements further help identify the location of sources. Moreover, it is generally believed that elevation perception is predominantly due to the complex structure of pinnae and the so-resulted spectral cues. HRTFs were developed and used mainly in auralization and VAS (Xie, 2013, Carlile, 1996), but their inference on the understanding of how human hearing works is profound.

It is common to describe locations around a listener using a spherical coordinate system as illustrated in Figure 6.1, with the centre of the head as origin. HRTFs for left and right ears H_L and H_R are defined as the transfer function between a source at a location in space and the receiving positions in the ears.

$$H_L(j\omega) = \frac{P_L(j\omega, r, \alpha, \delta)}{P_s(j\omega, r)} \tag{6.1}$$

and

$$H_R(j\omega) = \frac{P_R(j\omega, r, \alpha, \delta)}{P_s(j\omega, r)} \tag{6.2}$$

where H_L and H_R are a pair of HRTFs for left and right ears, P_L and P_R arc Fourier transforms of sound pressures measured in ears (strictly speaking at eardrums but often measured at alternative locations), α and δ are azimuth and elevation angles,

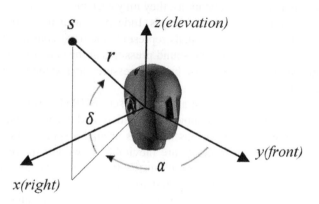

FIGURE 6.1 Commonly used coordinate system for HRTFs

and P_s is the Fourier transform source pressure, which is often measured at the centre of the head but without the presence of the head. Note that the front of the head is the y axis, where $\alpha = 0$, and rotates clockwise to the right onto the x axis. Elevation angle counts from x-y plane upwards as positive. Elevation is represented as z axis pointing upwards. The distance of the source to origin is r. It is worth noting that some authors may define the spherical coordinate systems differently.

The use of the HRTFs assumes that the transmission path from source to receiving position represents a linear time invariant (LTI) system, so the inverse Fourier transform of the HRTFs become the head related impulse responses (HRIRs). These show no difference to what was discussed in Chapter 3. Time limited HRIRs are often measured, and they give a pair of FIR filters that can be used for auralization or further calculation of the HRTFs via the Fourier transform. Minimum phase approximations of HRTFs are common practice using a pure delay and a minimum phase FIR, which will be discussed later in this section. On many application occasions, the distance between the source S and the ears are far greater than size of the head (when $r > 1$ m), and these are deemed as far field cases. It has been identified that in such far field conditions, the HRTFs are not particularly sensitive to distance.

6.1.2   HRTF Data

There are a number of ways one can obtain the HRTFs; each has its advantages and limitations.

1. **Measurement of HRTF of Human Subjects**

   The strict definition of the HRTFs is a pair of acoustic transfer functions describing the transmission path from a point in the spherical coordinate system to the two eardrums. However, HRTFs are rarely measured at the eardrums in human subjects for a number of reasons but mostly because of the potential risk of inserting the miniature probe microphone very close to the eardrums. Moreover, ear canals are very narrow spaces. The presence of the microphone changes the sound fields in the ear canals and the pressures on eardrums. Last but not least, HRTFs are of interest for various purposes. For certain applications, they may even be preferable to be measured at the entrance of the ear canals. Indeed, measuring at the ear canal entrances with blocked ear canals represents the most commonly adopted method. Moreover, should the sound pressures at the eardrums be needed, they can be estimated reliably by the pressures and velocities at the ear canal entrances.

   It is no easy task to obtain a complete set of HRTFs from human subjects. First, a population of subjects is needed, and second, sound sources need to be presented in various locations around a sphere. As a result, a large number of measurements are inevitable. If we are interested in far field HRTFs, distance from source to the listener may be discounted. To further mitigate the labour required to adjust the position of the sound source and automate the measurement process, the source can be mounted on a motorised cart mounted on a circular rail to transport the source for elevation position changes and a turntable used to move the subjects for azimuth angle changes. For far field measurements, a relatively small full frequency range loudspeaker is often suitable for playing the test stimuli.

The measurement techniques for HRTFs show no major differences to those discussed in Chapter 3. A sine sweep, MLS sequence, white noise, or other suitable broadband stimulus can be used. The HRIRs are typically measured in anechoic conditions. Some researchers measure them in spaces with no major reflective surfaces nearby and apply the time domain gating technique to emulate anechoic conditions. Since the HRIRs are much shorter (the effective duration is typically less than 2 ms) than room impulse responses, time variance during the measurement is less significant. MLS measurements are generally preferred to take advantage of their intrinsically high noise rejection property. Transducers, the loudspeaker and microphone involved in the measurement chain, have their impulse responses; in room acoustics, measurement transducer impulse responses are much shorter than those of the room and might be neglected. However, in HRTF measurements, transducer impulse responses are in the similar order of the HRIRs; they should be de-convolved out to eliminate large errors. When interpreting measured HRTFs, one needs to pay special attention to the measurement position, next to eardrum or at the entry of ear canal, open or blocked ear canals, etc. It is also important to look at statistical distribution and understand how individual HTRFs differ from the mean and medium of the dataset. To take full advantage of source localisation capability of human listening, individual HRTFs are needed. Arguably, this is the motivation of measuring the HRTFs of human subjects. Otherwise, a dummy can be used.

2. **Measurement Using a Dummy Head or HATS**

Measuring a complete set of HRTFs with human subjects means the subjects are inevitably exposed to the unpleasant testing signals for a prolonged period of time. Individual discrepancy in measures of outer ears and shapes of head mean that a good number of subjects are needed to identify the statistical distribution. To avoid these complexities, dummy heads or HATS are often used to study the averaged or representative HRTFs. The MIT Media Lab HRTF library is probably the most quoted universal HRIRs in the literature used by numerous researchers (Gardner & Martin, 1994). It was measured around a KEMAR dummy head over 700 spatial locations. The data can be used as FIR filters to auralize spatial sounds. When compared with testing using human subjects, however, the KEMAR HRIRS are known to have higher localisation errors in the elevation aspect but generally good results in the horizontal plane, apart from a certain level or front-back confusion (Moller et al., 1996).

3. **Estimation from Simple Models**

To study the HRTFs, one may wish to identify how the HRTFs or HRIRs vary when source location changes. One useful way to start the investigation is to use a simple model for the head, say a sphere without outer ear

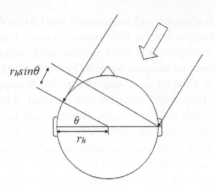

FIGURE 6.2   Simplified head model

modeling, as illustrated in Figure 6.2. This will enable us to roughly model ITDs and ILDs.

For example, using the simplified model, one can calculate the ITD by

$$ITD = \frac{r_h}{c}(\theta + \sin\theta) \qquad (6.3)$$

where c is the speed of sound, r_h is the radius of the head, and θ is the azimuth angle.

For more realistic head and ear shapes, analytical solutions tend to be less manageable, so numerical modeling can be used. Finite element modeling of a dummy or a model of a human subject obtained by 3D scanning may be used. Computational mesh is heuristically determined, Greens function evaluated, and FEM performed.

4. **HRTF Approximation**

In far field, when $r \geq 1$ m, HRTFs are not sensitive to distance. For spatial audio applications, far field data sets are often adequate. Since the distance from source to receiver is large, there is a significant delay before the test signal interacts with the ear and outer ear. That is to say, the impulse response is subject to distance and air attenuation only in the first part and then the actual head-related filtering processes. Visually, this can be seen in the top plot in Figure 6.4. The measured or modeled HRIRs represent linear but non-minimum phase systems. There are motivations to use shorter and minimum phase filters to approximate the original non-minimum phase HRIRs for computational efficiency and flexibility. In particular, the HRTF is often approximated by a pure delay type of all pass filter associated with the ITD and a minimum phase HRTF (Nam et al., 2008), as illustrated in Figure 6.3. If the head is considered to be a sphere, then the ITD can be calculated from Equation 6.3.

FIGURE 6.3   Delay and minimum phase approximation

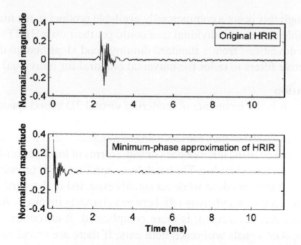

FIGURE 6.4   An example of measured HRIR and minimum phase approximation

For binaural synthesis, HRTFs are commonly implemented as pure delays followed by minimum-phase systems. Here, the minimum-phase nature of HRTFs is studied. The cross-coherence between minimum-phase and unprocessed measured HRTFs was seen to be greater than 0.9 for a vast majority of the HRTFs, and was rarely below 0.8. Non-minimum-phase filter components resulting in reduced cross-coherence appeared in frontal and ipsilateral directions. The excess group delay indicates that these non-minimum-phase components are associated with regions of moderate HRTF energy. Other regions of excess phase correspond to high-frequency spectral nulls, and have little effect on cross-coherence.

As an example, one can use an FIR-based adaptive filter to identify the minimum phase part of an HRTF, by removing the leading zero (i.e. the transmission delay found in measured HRTFs).

6.1.3   APPLICATION SCENARIOS

1. **Binaural Recording**

The underpinning theory of sound source localisation capability of binaural hearing is the ITD, ILD, and spectral cues as they arrive at the two eardrums. One established and also the simplest way to record and reproduce spatial sound is to pick up the signals in-situ and in-ear next to eardrums and play them back at the exact same locations. This is known as binaural recording and playback technique. It works almost perfectly provided that (1) the recording is individualised, i.e. playing back the signals in the same ears as they were recorded, and (2) there is no head movement during the recording and playback phases.

Binaural recordings using universal head models, such as commercially available dummy heads for listeners to enjoy spatial sound concerts, have been experimented with by some broadcasting companies. Even though the results are not perfect when compared with the use of individualised HRTFs, these universal live binaural recordings are entertaining.

Although this is not a commercially available product in the marketplace, it is possible that each individual user could get their own HRTFs measured and have deviations from a standard dummy head identified, to derive a set of corrective filters to tailor the universal binaural for individual users.

2. Auralization

Auralization is the technique of rendering virtual 3D sound spaces. In most cases, auralization puts emphasis on the means to create or synthesise spatial sound virtually, rather than the capturing of a spatial sound field. Auralization is often achieved using microphones or some forms of loudspeaker arrays, ideally in an anechoic chamber. The headphone auralization is probably the simplest, since it only needs to work out signals expected in listeners' ears; there is no room acoustics and cross-talk between channels involved. Auralization in an enclosure, i.e. a room, is more complicated. It encounters cross-talk, i.e. all speaker signals will reach both ears. If there are sound reflections in the enclosure, false virtual sources may occur. Various techniques have been developed to rectify these problems; amongst them, the wavefield synthesis represents the most sophisticated approach to the virtual sound spaces.

Headphone auralization can be achieved using HRTFs. Given source locations and corresponding HRTFs, the headphone signals can be derived by convolving the source with the corresponding HRIRs. Since the linearity is assumed and generally satisfied, multiple sources in different locations are represented by the summation of the convolved individual signals at the receiving points. More specifically, for n point sources $s_1(t)$ to $s_n(t)$, if the HRIR pairs for left and right ears corresponding to the locations of the sources are $[h_{1L}(t), h_{1R}(t)]$ to $[h_{nL}(t), h_{nR}(t)]$, then the left and right headphone signals $s_L(t)$ and $s_R(t)$ can be determined by

$$s_L(t) = s_1(t) \otimes h_{1L}, + s_2(t) \otimes h_{2L}(t) + s_3(t) \otimes h_{3L}(t) + \ldots + s_n(t) \otimes h_{nL}(t) \quad (6.4)$$

$$s_R(t) = s_1(t) \otimes h_{1R}, + s_2(t) \otimes h_{2R}(t) + s_3(t) \otimes h_{3R}(t) + \ldots + s_n(t) \otimes h_{nR}(t) \quad (6.5)$$

3. Signal Processing for Transaural Audio Rendering

Most commonplace stereo recordings are produced for loudspeaker rendering. When these were rendered through headphones, unnatural spatial sensations can occur due to the lack of cross-feed, i.e. the sound propagates from right speaker to the left ear and left speaker to the right ear. A cross-feed processing stage to derive the signals for headphones can mitigate the problem. Conversely, rendering binaural recordings through a pair of stereo speakers justifies the use of cross-talk cancellation. There have been many attempts to improve perceived sound quality and mitigate the problems due to mismatched playback platforms, which are termed transaural rendering. We discuss in this section how HRTF and DSP can be used to solve the problems.

Rendering typical stereo recordings intended for loudspeaker playback over headphones are known to be problematic (Linkwitz, 1971), and this is due mainly to the lack of cross feeding from left channels to the right ear and right channels to the left ear of a listener. An ideal processing unit to translate stereophonic to headphone signals should incorporate ITD, ILD, and spectral cues resulting from scattering diffraction and reflections of the pinnae, head, and torso. The use of simpler cross-feed techniques to improve headphone listening experience is not new. Analogue cross feeders have been in use in high-end headphone amplifiers since the same era when stereophonic recordings became prevalent, e.g. the well-known Linkwitz cross-feed circuit (Linkwitz, 1971). These circuits add attenuated and low pass filtered signals to the cross channels. Typically, the low pass filter has a cut-off frequency at about 700 Hz and a decay scope of 6 dB/oct. This means that the ILD and some coarsely approximated spectral cues are taken into account but the amount of ITD can hardly be compensated for with analogue circuits. Modern DSP cross-feed filters, often as software plug-ins, are potentially more accurate since it is possible to incorporate subtle details of HRTFs including sufficient delay time to model the ITD. As discussed in the previous two sub-sections, both headphones and cross feeding require personalisation to unlock their full potentials. DSP-based cross-feed filters well position themselves to these tasks. On the other hand, a universal type of modern cross filter would be beneficial for general users or as a starting point for further fine tuning and personalisation.

Cross-feed filters may be viewed as filters that correctly map two loud-speaker signals onto two headphone signals. If speakers are viewed as point sources and listening position is assumed to be ideal, which users always wish to be, then one pair of symmetric HRTFs, say at $r = 2$ to 3m, $\delta = 0$, and $\alpha = +/-30$ degrees, can be used to map speaker signals onto headphone ones. The azimuth angle α here is defined to be zero when facing the front, minus for left, and plus for right. Since $r > 1$ m, far field HRTFs are considered. The transfer matrix H relates the left and right headphone and left and right speaker signals S_{hL}, S_{hR}, S_{sL}, and S_{sR} respectively, all in the Fourier transform frequency domain.

$$\begin{bmatrix} S_{hL}(j\omega) \\ S_{hR}(j\omega) \end{bmatrix} = H \cdot \begin{bmatrix} S_{sL}(j\omega) \\ S_{sR}(j\omega) \end{bmatrix} \tag{6.6}$$

The transfer matrix H for a typical stereophonic set-up is

$$H = \begin{bmatrix} H_L(j\omega, -30°) & H_L(j\omega, +30°) \\ H_R(j\omega, -30°) & H_R(j\omega, +30°) \end{bmatrix} \tag{6.7}$$

where H_L and H_R are HRTFs for left and right ears. If only cross feed is considered, H can be simplified to

$$H' = \begin{bmatrix} 1 & H_L(j\omega, +30°) \\ H_R(j\omega, -30°) & 1 \end{bmatrix} \tag{6.8}$$

If HRIRs are known (e.g. from the MIT KEMAR HRIR data set), they can be used to derive headphone signals directly by convolutions in the time domain. Two methods have been investigated in this paper to determine the universal cross-feed filters. The first one used the MIT KEMAR HRIRs auralization to render the speaker signals on headphones. The second started with the aid of an adjustable HRTF-based "virtual listening room via headphone" plug-in. Subjects were asked to set the parameters to their preference. The median of settings of all subjects over all audio samples were taken and the associated transfer functions were measured to determine the universal cross-feed processing.

There are motivations to make binaural recordings of live events using dummy heads, since they offer 3D sound over existing 2-channel delivery systems if headphones are used by the listeners, and preserve useful spatial information for possible separation and re-mixing in the future when relevant techniques become mature. There is, again, a compatibility issue when the binaural signal is rendered through the stereo pair of loudspeakers, simply because they impose extra unnecessary cross feed. The traditional method of playing back binaural recordings was to place a large baffle between the left and right speakers to provide a certain level of separation, but this is less effective in lower frequencies. Furthermore, large baffles are, on many occasions, not practical logistically.

Figure 6.5 illustrated the differences between loudspeaker rendering and headphone rendering. Since the sound from one channel (left or right) is also transmitted through the air to the contralateral ear, filtering the original signals to cancel the cross talk is necessary and possible (Li, 2015, Cheng & Wakefield, 2001). The principle of cross-talk cancellation can be described in matrix notation. In the following analysis, we will use two ears and two loudspeakers with the amplitudes Y_L and Y_R for the left and

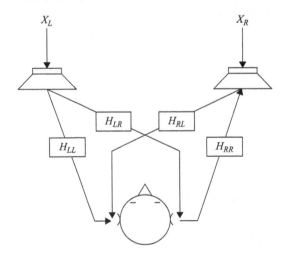

FIGURE 6.5   Relations between headphone and speaker signals

the right channel, respectively. There is a relation between outputs at both channels and the desired amplitudes at our ears, X_L and X_R:

Equation 6.5 can be further expanded to

$$
\begin{bmatrix} S_{hL}(j\omega) \\ S_{hR}(j\omega) \end{bmatrix} = \begin{bmatrix} H_{LL}(j\omega) & H_{LR}(j\omega) \\ H_{RL}(j\omega) & H_{RR}(j\omega) \end{bmatrix} \cdot \begin{bmatrix} S_{sL}(j\omega) \\ S_{sR}(j\omega) \end{bmatrix}
\tag{6.9}
$$

Comparisons between universal HRTFs and individualised HRTFs have been reported by a number of authors, e.g. results from (Moller et al., 1996). These results suggested the necessity of adopting individualised HRTFs for accurate source localisation. Väljamäe et al. (2004) compared the individualised and generic HRTFs in video games (i.e. audio with visual cue) and showed significant advantages of using individualised HRTFs. Nonetheless, measuring individual HRTFs is complicated and is not practicable for the majority of headphone consumers. For cross-feed filters, there may not be as accurate a source localisation requirement as that required for gaming and the VAS. It is, therefore, postulated that universal cross feed might still offer improved listening experience. Subjective tests have, therefore, been used to identify if, and to what extent, these two proposed universal cross-feed filters can improve listening experience.

KEMAR dummy head and torso simulators have been used for hearing-related research since the 1970s. A large number of studies into the universal or generic HRTFs were carried out using the KEMAR simulators. The most-broadly quoted dataset is probably the MIT HRIR library (Gardner & Martin, 1994). A straightforward set of cross-feed filters was implemented with the MIT KEMAR HRIR data for +/− 30 degrees and far field (2 m). Straightforward convolution of the speaker signals and the HRIRs yields signals for headphones. With this method, there is no parameter to be adjusted or averaged, since the KEMAR emulator is assumed to be the generic model for all. Subjective tests were carried out. Subjects perceived improvement of listening experience but some common artifacts of headphone rendering as well. This is not surprising as other authors reported similar findings when using non-personalised HRIRs (Gardner & Martin, 1994, Väljamäe et al., 2004). Results will be discussed in the next section.

The HRTFs were measured for several subjects by Dr. John Middlebrooks at the Kresge Hearing Research Institute at the University of Michigan, using in-ear, probe-tube microphones. Measurements were taken in an anechoic chamber using the method described in 1.2.1 above. Complementary Golay codes were used as the stimulus signals and were presented from a loudspeaker approximately 1.5 m from the subjects' heads. Left and right ear magnitude responses were measured at 400 different azimuth-elevation locations. Although irregularly spaced, these locations are roughly 10° to 15° apart in the azimuth and elevation directions. The sampling rate was 50 kHz, the resolution of the data taken was 16 bits, and a 512-point FFT was used to compute the frequency response at each location.

There is no major difference between the measurement of HRTFs and other transfer functions discussed in Chapter 3. One may use sine sweep, white, or pink noise, or MLS or narrow pulses. Their advantages and disadvantages remain the same. However, there are some practical considerations in HRTF measurements.

1. By definition, common HRTFs are the HRTFs measured in an anechoic chamber. Time-domain gating may be used if anechoic conditions are not available.
2. The location of the transducer is at the eardrum position for the dummy for human subjects; typically it is set to be 1 to 2 mm from the eardrum. This poses a number of problems. Miniature microphones possible to be placed in-ear can only provide limited sensitivity and frequency responses. Placing objects very close to eardrums poses the risk of harming subjects. As a result, the common practice is to measure both the pressure and velocity at the entrance of the ear canals (Moller et al, 1996).
3. To save time, many authors prefer the use of MLS for measurement of HRTFs since time variance is not significant, especially when measurement is with a dummy head.

6.2 MICROPHONE ARRAYS AND DELAY-SUM BEAMFORMERS

A microphone array is a set of acoustic transducers that employs two or more microphones. The key technology or the main purpose of using a microphone array is beamforming, which means to synthesise special directivity patterns, typically hyper directional ones. This enables the array of microphones to work jointly and focus on a particular direction in which the source of interest is located.

The ideas of transmitter and receiver arrays started from antenna arrays for radars to enhance transmission and reception. Traditional radar systems typically use directional antennas, e.g. parabolic dishes, which can be mechanically steered in elevation and rotated constantly to scan the space and identify targets. To achieve high signal-to-noise ratios and close-to-continuous monitoring, large dishes and a high rotation speed are needed. These are limited by logistical and technical constraints. The phased-array antennas were developed to address these problems. In such an array, and taking the receiving mode as an example, multiple antennas are used, a phase shift is applied to the output of each antenna input/output, and then summed up as the output. This is the basic array that is still broadly used today, known as the delay-and-sum beamformer. In addition to radar and other antennas, arrays have gained broad applications in sensors and transducers, particularly in acoustics, e.g. sonar, ultrasonic arrays, speaker and microphone arrays, and acoustic cameras. In a broader sense, since the microphone array beamformers are used for spatial selectivity, microphone array signal processing is a time and spatial signal filtering problem. There are many applications of microphone arrays, together with suitable signal processing methods. These include, but are not limited to, hyper directional virtual microphones, signal-to-noise ratio enhancement, direction of arrival (DOA) detection, and source separation and localisation. From an application point of view,

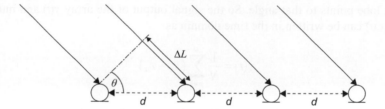

FIGURE 6.6   A uniform linear array (ULA)

beamforming is a method to discriminate between source signals using spatial information of the sources. Directivity shaping and beam steering are the main themes of beamforming for arrays. Depending upon the operational modes, arrays can be categorised as fixed ones and adaptive ones. The former use a fixed set of parameters that do not change during the operation phase. The latter adopts adaptive algorithms and therefore the parameters are updated during the operation on the fly to minimise a purposely chosen cost function. A simple example is given to show some basic ideas of microphone arrays and array signal processing.

Figure 6.6 shows four consecutive microphones of a uniform linear array (ULA) with N microphones uniformly placed with a distance between two adjacent microphones. When the microphone array picks up a signal coming from an angle θ, every consecutive microphone will experience a time delay due the extra distance ΔL that the sound travelled.

The time delay for the consecutive microphone to receive the same signal τ can be determined for d, θ, and sound speed c by

$$\tau = \frac{d}{c}\sin\theta \qquad (6.10)$$

since the microphone array is not a continuous linear array but a discrete one. This is, effectively, spatial sampling. To avoid spatial aliasing, d should be shorter than half of the wavelength λ of the signal, which then determines the useful frequency upper limit of the array.

$$d < \frac{\lambda}{2} \qquad (6.11)$$

If microphone signals are in phase, they are constructive; if signals are out phase, they cancel out. Different phases in microphones will cause interference, and by processing the delayed signals, various directivity patterns can be achieved. This is essentially how array works. When we wish to steer the array to a specific angle, we can change the phase weightings applied to microphone signals from each channel so that a main lobe points to that angle. The simplest way to achieved this is the use of delay and sum (DAS), or delay-sum, beamforming. The terminology is "delay and sum beamforming" because in the time domain, microphone inputs are first delayed and then summed to give a single output. Properly delaying the signals from consecutive microphones (except for the last microphone, as it is often used as a reference) is done to ensure the signals from the selected angle interact constructively, so that a main lobe points to that angle. So the signal output of the array $y(t)$ and multiple inputs $x_i(t)$ can be written in the time domain as

$$y(t) = \frac{1}{N} \sum_{n=1}^{N} (x_n - \tau_n) \tag{6.11}$$

It can be proven that to steer the array to the direction with an angle θ, the mth microphone in the line delay with reference to the first microphone should be

$$\tau_m = \frac{(m-1)d}{c} \cos\theta \tag{6.12}$$

Apparently, to achieve a desirable directivity pattern, delays need to be correctly applied and the delays, i.e. phase changes, are not just a simple function of angle but a function of frequency as well. This is the reason why signal processing algorithms are needed.

For the mth channel, the frequency domain weighting should be

$$w_m(f) = \frac{1}{N} e^{-j\frac{2\pi f(m-1)d\cos\theta}{c}} \tag{6.13}$$

and, thus, the system output becomes

$$y(f) = \frac{1}{N} \sum_{m=1}^{N} x_m(f) e^{-j\frac{2\pi f(m-1)d\cos\theta}{c}} \tag{6.14}$$

Since the delay-sum beamformer is, in fact, a frequency domain weighting to all input channels, it becomes easier to be implemented in the frequency domain. Figure 6.7 shows a commonly used classical delay-sum beamforming algorithm.

There are beamforming algorithms; some are based on fixed DSP algorithms, while others rely on adaptive methods.

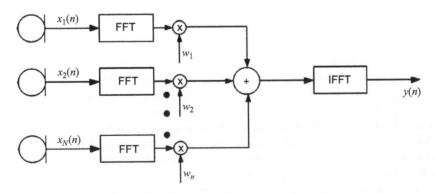

FIGURE 6.7 Implementation of a classical sum-delay beamformer in the frequency domain

SUMMARY

Binaural nature means that audio can ideally encode and reproduce spatial information of sound. Spatial audio extends monaural audio to a new horizon and represents a step towards more naturalistic and informative audio. On the other hand, acoustics as a branch of physical science can make use of acoustic signals to localise the sources, and acquire spatial information of vibration and sound propagation for many important and interesting applications. Spatial audio is in a dynamic and rapid development and deployment phase today thanks to the enabling modern technologies offering affordable solutions. There exist many systems, schemes, or techniques for spatial audio. The lack of a unique and clear market winner, to some extent, contributes to the diversity. Microphone arrays show a lot of potential, not only in acoustical measurements but audio applications as well. They can take the simplest line formats or much more complicated shapes to offer source localisation, separation, or even acoustic imaging. Spatial audio and microphone array technologies are specialist fields, which cannot be summarised in a chapter of a signal processing book. This chapter has very briefly discussed, from a few examples, how signal processing plays a vitally important role in spatial audio and microphone arrays. Arguably, spatial audio and microphone array technologies are predominant signal processing problems.

REFERENCES

Bai M. R and Lin, C. "Microphone Array Signal Processing with Application in Three-Dimensional Spatial Hearing," *Journal of the Acoustical Society of America*, 2005, Vol. 117 (4) Pt 1, pp. 2112–2121.

Benesty, J., Chen, J., Huang, Y. A., and Dmochowski, J. "On Microphone-Array Beamforming from a MIMO Acoustic Signal Processing Perspective," *IEEE Transactions on Audio, Speech, and Language Processing*, 2007, Vol. 15, No. 3, pp 1053–1064.

Cheng C. I. and Wakefield, G. H. "Introduction to Head-Related Transfer Functions (HRTFs): Representation of HRTFs in Time, Frequency, and Space," *Journal of the Audio Engineering Society*, 2001, Vol. 49, No. 4, pp. 231–249.

Gardner, N. and Martin, K. (1994) "HRTF Measurements of a KEMAR Dummy-Head Microphone," MIT Media Lab Perceptual Computing—Technical Report #280.

Li, F. F. (2015) "Improving Headphone User Experience in Ubiquitous Multimedia Content Consumption: A Universal Cross-feed Filter," *Proceedings of the IEEE International Symposium on Broadband Multimedia Systems and Broadcasting*, pp. 1–6.

Linkwitz, S. "Improved Headphone Listening—Build a Stereo-Crossfeed Circuit," *Audio*, 1971, Dec., p. 43.

Moller, H., Sorensen, M. F., Jensen, C. B., and Hammershoi, D. "Binaural Technique: Do We Need Individual Recordings?" *Journal of the Audio Engineering Society*, 1996, Vol. 44, No. 6, pp. 451–469.

Nam, J., Kolar, M. A., and Abel, J. S. (2008) "On the Minimum-Phase Nature of Head-related Transfer Functions," *AES 125th Convention*, Paper Number: 7546.

Shinn-Cunningham, B. (1998) "Application of Virtual Auditory Displays," *Proceedings of the 20th International Conference of IEEE Engineering in Biology and Medicine Society*, Vol. 20, No. 3, pp. 1105–1108.

Väljamäe, A., Larsson, P., Västfjäll, D., and Kleiner, M. (2004) "Auditory Presence, Individualized Head-Related Transfer Functions, and Illusory Ego-Motion in Virtual Environments," *Proceedings of the Seventh Annual Workshop*, pp. 141–147, Valencia.
Yu, G., Xie, B., and Chen, X. "Analysis on Minimum-Phase Characteristics of Measured Head-Related Transfer Functions Affected by Sound Source Responses," *Computers and Electrical Engineering*, 2012, Vol. 38, pp. 45–51.

BIBLIOGRAPHY AND EXTENDED READING

Xie, B. (2013) *Head-Related Transfer Function and Virtual Auditory Display*, J. Ross Publishing.
Rumsey, F. (2001) *Spatial Audio*, Focal Press.
Bai, M. R., Jeong-Guon Ih, J-G., and Benesty, J. (2014) *Acoustic Array Systems: Theory, Implementation, and Application*, Wiley-Blackwell.
Benesty, J. (2016) *Fundamentals of Differential Beamforming*, Springer.
Carlile, S. (1996) *Virtual Auditory Space: Generation and Applications*, Springer-Verlag.

EXPLORATION

MINI PROJECT IDEAS

Cross-feed filters for headphones may be derived from the HRTFs of the specific location of a typical set up of a stereophonic audio system (e.g. +/–30 degrees of azimuth angles, 2-3 metres of listening distance, and 0 degree of elevation). As a mini project, you can explore the use of the universal HRTFs, such as these specified in the MIT KEMAR dummy head dataset, to design a cross-feed filter on the MATLAB® platform. Use the cross-feed filter to pre-process a standard stereophonic recording and convert it to signals more suitable for headphone rendering. Listen through the headphone to the original standard stereophonic and processed signals; do you notice any improvement by cross-feeding?

7 Adaptive Filters

Traditional filters manipulate signals in a predetermined manner. For instance, a speech signal recorded on an old cassette recorder is contaminated by a hissing noise mainly in higher frequencies above 8 kHz and a rumbling mechanical noise due to the tape transport mechanism in lower frequencies below 100 Hz. A band pass filter with a pass band from 125 to 6300 Hz may be used to improve the quality of the speech recording, since it removes the noises outside the frequency band of speech and keeps the speech signal intact. Based on the knowledge about the noise and wanted signals, design specifications can be drawn up and filter coefficients determined using a suitable design technique, as discussed in previous chapters. Once the filter is designed and implemented, its coefficients are fixed. However, in many other cases, the types of noises or interferences are unknown, and they may also be time-varying. Fixed filters can hardly achieve optimal results. A new type of filter that will continuously track the changes of signal characteristics and/or environment, and adjust its parameters (coefficients) to adapt itself to such changes is needed. Adaptive filters have been proposed and developed to address this issue.

An adaptive filter is often built around a linear FIR or IIR filter, but the filter coefficients are iteratively updated. To design an adaptive filter, a cost function is defined, which sets the criteria of optimisation. A suitable optimisation algorithm is then used to minimise the cost by iteratively updating the filter coefficients. Thus, an adaptive filter tracks the changes of environment, changes the filter transfer function accordingly, and achieves the optimal performance. The most popular cost function is the mean square error of the signal, and the associated family of adaptive filter algorithms are called least mean square (LMS) algorithms. Another type of established adaptive filter algorithm is the recursive least square (RLS) filter, in which the weighted linear squares cost function is optimised. It is worth noting that LMS algorithms assume the input signal is scholastic, while RLS takes the input signal as deterministic. Figure 7.1 shows simplified general block diagrams for LMS and RLS adaptive filters.

The LMS filters are the most widespread type of adaptive filters. Both LMS and RLS filters employ either an FIR or IIR as their core filtering block. The parameters, i.e. coefficients, are updated and, hence, the transfer function is changed continuously by the adapting algorithms. In an LMS filter, the output of the filter is checked against the desired signal to yield the error signal. The mean square errors are used as a criterion for the optimisation. The adapting algorithms adjust the filter parameters to minimise the mean square errors. In this way, the filter is adapted to achieve optimal performance, i.e. the output will closely approximate the desired one in a least mean square sense. In the LMS filter, the input signal is assumed to be stochastic and is not involved in determining the adapting strategy. In the RLS filter, both the input signal and the error signals are used to determine the how the filter is to be updated. (The square errors are weighted by input.) In RLS filters, inputs are assumed to be deterministic.

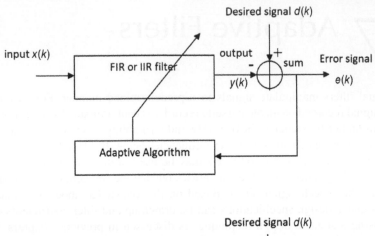

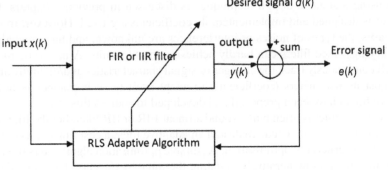

FIGURE 7.1   LMS (top) and RLS (bottom) adaptive filters

To evaluate cost functions, adaptive filters utilise a closed loop feedback structure. This adds constraints to the design to avoid instability problems. How stable
an adaptive filter is, how quickly it can converge, and how accurately it can replicate
the wanted signals are the three major concerns in assessing the performance of
such filters.

Computer hardware and dedicated DSP processors have gained substantial
advances; today, adaptive filters are commonly used in mobile phones, hearing
aids, and many other consumer electronics devices. General applications include
system identification, noise or unwanted signal component removal, prediction,
and many others.

Linear adaptive filters are built around linear filters. This means that at any
single sampling point in time, the input-output mapping relationship of such filters is linear, i.e. superposition theorem applies. Over time, the parameters of
such linear filters can change as the input signal and environment change. In
this sense, the behaviour of a linear adaptive filter can be non-linear and time
variant. From this point of view, adaptive filters extend the linear time invariant
filters to applications in which non-linear filtering and time variance need to be
considered. Moreover, the inner filter of an adaptive filter is not restricted to any
particular type.

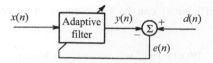

FIGURE 7.2   A general model for adaptive filters

7.1 GENERAL MODEL OF LMS ADAPTIVE FILTERS

In this and the next three sections, emphasis will be placed on the LMS type of adaptive algorithms. Figure 7.2 shows a further simplified universal structure of LMS filters, in which $x(n)$ is the input to the filter, $y(n)$ the output of the filter, $d(n)$ is the designed outcome, and $e(n)$ is the error.

Clearly, the error signal is

$$e(n) = y(n) - d(n) \tag{7.1}$$

In this general model, how ensemble errors are calculated and how adaptation is performed are not specified. These essential components in the adaptive filter are all implied by the arrow across the filter.

On seeing this block diagram, one might ask the questions: "How do we know $d(n)$?" and "If we knew the desired outcome or response, why would the filter really be needed?" These will be discussed in the next section. In essence, to make use of the attractive features of adaptive filters, problems need to be cast in this model. However, this does not necessarily mean the desired response is of direct interest. In fact, sometimes the interests are in the filter coefficients, or even the error signals! There are certainly cases in which the desired response is what is of interest. In such cases, an estimate or a reference signal might be needed. Indeed, the design of an adaptive filter starts from translating the problem into the adaptive filter model; the rest of the task will essentially be the development of suitable optimisation algorithms that determine update regimes.

Consider a transversal FIR filter as shown in Figure 7.3.

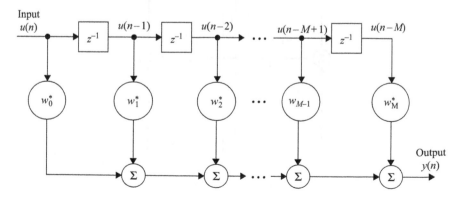

FIGURE 7.3   Structure of a transversal filter

Its parameters (coefficients) can be represented by a vector $W(n)$ that is the impulse response

$$W(n) = [w_0, \ w_1, \ w_2, \ w_3, \ \cdots \ w_{N-1}]^T \tag{7.2}$$

So

$$y(n) = \sum_{i=0}^{N-1} w_i(n)x(n-i) \tag{7.3}$$

$$= W^T(n)X(n)$$

The use of vectors to represent signals and filter coefficients makes the difference equation clearer and neater. These notations are often used in deriving adaptive filters.

Consider an IIR filter as illustrated in Figure 7.4.

The output becomes

$$y(n) = \sum_{\Sigma=1}^{N} a_i(n)y(n-i) + \sum_{\Sigma=1}^{N} b_j(n)y(n-j) \tag{7.4}$$

If an input-output relation expression consistent with the one used for FIR is sought, then the filter vector and input vector can be arranged as

$$y(n) = w^T(n)P(n) \tag{7.5}$$

where

$$W(n) = \left[a_1(n)a_2(n) \cdots a_N(n)b_0(n)b_1(n) \cdots b_N(n) \right]^T$$

$$U(n) = [y(n-1)y(n-2) \cdots y(n-N)x(n)x(n-1) \cdots x(n-N)]^T$$

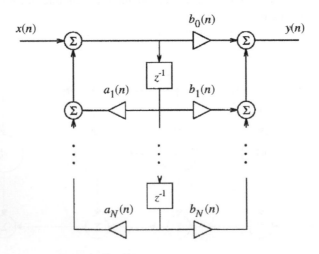

FIGURE 7.4   An IIR filter

7.2 FOUR GENERIC TYPES OF ADAPTIVE FILTERS

Adaptive filters have found numerous applications in diverse fields. Some applications share certain features in common and can be treated as one specific type of application. The categorisation below is based on the nature of the tasks and the system configurations. It is important to note that in various adaptive filter configurations, the direct inputs to the adaptive filters themselves may not necessarily be in the inputs of the whole system; similarly, the outputs of the adaptive filters may not be the ones of actual interest. The four generic types of adaptive filters may be viewed as building blocks for more complicated adaptive filters or systems, which combine two or even more of these blocks.

7.2.1 SYSTEM IDENTIFICATION

One common application of adaptive filters is to use such filters to "identify" an unknown system, such as the response of an unknown communications channel or the transfer function of a loudspeaker. More broadly, this can be viewed as the use of an adaptive filter to approximate an unknown filter. For system identification applications, the common system configuration is illustrated in Figure 7.5, in which the unknown system is connected in parallel with the adaptive filter. The aim is to update the filter coefficients so that the adaptive filter can behave in a manner as similar as possible to the unknown system when they are exposed to all sorts of possible input signals $x(k)$. Here, $x(k)$, which is applied as inputs to both the unknown system and the adaptive filter, is a probe stimulus and is, typically, a broadband noise that covers the frequency range of interest. When this is translated into an adaptive filter model, the error signal $e(k)$ is the difference between the output of the adaptive filter and the unknown system, and the objective becomes the minimisation of the error function $e(k)$ in a least mean square sense. It is worth noting that the accuracy of such system identification is limited by the inner filter of the adaptive filter. If the unknown system is complicated but the inner filter of the adaptive filter has only a few taps, the accuracy of identification will be limited by the number of taps. This may not be a disadvantage in certain applications. One may use simple filters to approximate complicated input-output mapping

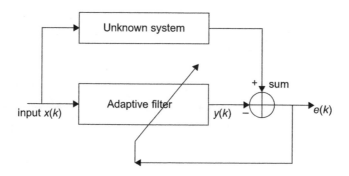

FIGURE 7.5   An adaptive filter for system identification

relationships. It is clear that, for this type of application, the output of the filter is not of direct interest. What we try to find is the internal parameters of the adaptive filter, i.e. the coefficients.

7.2.2 INVERSE MODELLING

Finding the inverse of an unknown filter or system is sometimes needed. One may think that it can be done by finding the unknown filter using the model discussed above first and then find the inverse transfer function. To find an inverse transfer function for a system is not always an easy task. An example is if a system is non-minimum phase, i.e. the zeros of the transfer function are outside the unit circle. The strict inverse of such a system does not converge, since the zeros will become poles in inverse transfer function resulting in an unstable system. Nevertheless, approximating the inverse with a causal and stable filter is possible. As illustrated in Figure 7.6, the unknown system is arranged in series with an adaptive filter with a causal and stable inner filter structure, say, an FIR. The filter adapts to become an approximate of the inverse of the unknown system as $e(k)$ is minimised. Since both the unknown and the adaptive filter cause delays to the signals passing through them, a delay unit as shown on top in the block diagram in Figure 7.6 is, therefore, needed to keep the data at the summation synchronised so that an effective comparison can be made to determine the error signal $e(k)$. Again, a broadband probe stimulus is needed as the source signal $s(k)$, but $s(k)$ is not applied directly as the input to the adaptive filter $x(k)$; rather, it is the input of the whole system, as illustrated in Figure 7.6. Similar to system identification applications, the interest here is not the output of the adaptive filter $y(k)$, but its coefficients.

7.2.3 NOISE OR INTERFERENCE CANCELLATION

The most common applications of adaptive filters are probably adaptive noise removal or reduction, more often referred to as "noise cancellation" algorithms. In adaptive noise cancellation, adaptive filters track and mitigate varying noises from a signal in real time through continuous updating. Figure 7.7 shows a typical configuration of adaptive filter for such applications.

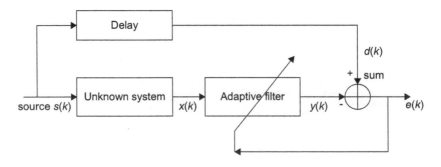

FIGURE 7.6   System inverse problems

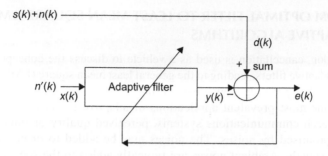

FIGURE 7.7   Adaptive filter for noise cancellation

A reference signal $n'(k)$, which correlates with noise, is fed to the adaptive filter as its input $x(k)$. The signal to be de-noised, which contains the source signal $s(k)$ and noise signal $n(k)$, is viewed as the desired signal $d(n)$ according to the general model shown in Figure 7.2. Since the input $x(t)$ to the filter is correlated to the noise in the wanted signal, the filter adapts its coefficients to continuously reduce the difference between $y(k)$ and $d(k)$. The error signal $e(k)$ converges to the wanted signal; it becomes the actual output in real applications. Clearly, the adaptive filter output $y(k)$ converges to the noise; the adaptive filter here functions as a noise estimator.

7.2.4   LINEAR PREDICTION

Prediction means to estimate the future value of a signal based upon the information about its values in the past and at present. To predict, some essential assumptions must be made. Tangible prediction is only possible when these conditions are met or approximately obeyed. Typical assumptions necessary for linear predictions are: (1) the signal is generally periodic, and (2) the signal varies slowly over time.

Figure 7.8 shows an adaptive configuration for linear prediction.

If the observation $s(k)$ is a periodic signal contaminated by some random noises and the adaptive filter is long enough to cover one period, such an adaptive filter with a delay can predict what is the true value in the future. The prediction here, essentially, is to remove random noise in a fundamentally periodic signal to recover its original periodicity nature, thus, giving prediction.

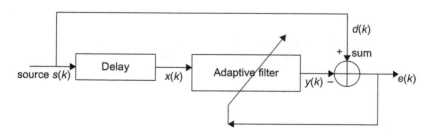

FIGURE 7.8   An adaptive filter for prediction

7.3 FROM OPTIMAL FILTER TO LEAST MEAN SQUARE (LMS) ADAPTIVE ALGORITHMS

In this section, cancellation is used as a vehicle to discuss the concept of optimal filters and adaptive filters, leading to the general least mean square (LMS) algorithm, the most popular approach to adaptive filtering.

One of the most prevalent applications of adaptive filters is noise cancellation. In speech communications systems, perceived quality or intelligibility is often compromised by noises. The noises may be added to or convolved with the speech signals. Ambient noises are typically added to the speech, while the impulse responses of transmission channels (acoustic and electronic ones) are convolved with the speech signal. It is possible to remove some of the additive noise using adaptive filters; convolved noises can be more difficult to tackle depending on the complexity of the impulse responses. We will discuss the cancellation of additive noises.

7.3.1 CONCEPT OF OPTIMAL FILTERS

Consider the system illustrated in Figure 7.9. The speech signal s is corrupted by the additive noise signal n at the first summing node, generating the observable signal d. At the second summing node, a signal y is subtracted from d. The result of this subtraction gives the error signal, e.

- If signal y is a copy of the signal n, then the noise corruption on the signal s is removed, $e = s$.
- If y is a reasonable approximation of n, then noise contamination is mitigated, $e \approx s$.
- If y is largely uncorrelated with n, then the second summing node represents an additional source of noise, further corrupting the speech component in e.

The cancelling signal y is derived by filtering operations (through the filter block W, which is not predetermined but can be optimised following some criteria) on the reference signal x. We now consider the necessary relationship between the signals and the optimal configuration of the filter that results in attenuation of the noise component of d.

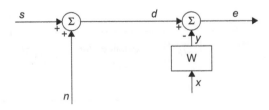

FIGURE 7.9   The fundamental structure of a noise-cancelling system

The error signal, e, can be written as:

$$e = s + n - Wx \qquad (7.6)$$

Without restricting types of filters and signal representation, all the signal variables and the filter response are considered as complex functions of frequency here. The magnitude squared error is then:

$$|e|^2 = |s + n|^2 - |W|^2 |x|^2 - 2\,\mathrm{Re}\left[(s+n)(Wx)^*\right] \qquad (7.7)$$

in which the superscripted asterisk denotes the complex conjugate and Re denotes real part. In order that the analysis can deal with the non-deterministic signal types typical of most audio signals, we apply some averaging of the mean squared error signal, using the expected value operator:

$$E\left[|e|^2\right] = E\left[|s|^2\right] + E\left[|n|^2\right] + 2E\left[sn^*\right] + |W|^2 E\left[|x|^2\right] - 2\,\mathrm{Re}\left[WE\left[sx^*\right]\right] - 2\,\mathrm{Re}\left[WE\left[nx^*\right]\right] \qquad (7.8)$$

We now make some assumptions about the statistical relationship between the signals:

1. The noise n is uncorrelated with the speech s.
2. The reference x is correlated with the noise n (and so, by 1); is uncorrelated with the speech s.

Under these assumptions, Equation 7.8 simplifies (since the averaged product of uncorrelated signals is zero) to:

$$E\left[|e|^2\right] = E\left[|s|^2\right] + E\left[|n|^2\right] + |W|^2 E\left[|x|^2\right] - 2\,\mathrm{Re}\left[WE\left[nx^*\right]\right] \qquad (7.9)$$

Notice that Equation 7.9 is a positive definite quadratic function of W—the mean square error will have a unique minimum value when the (complex) filter W is correctly adjusted to W_{opt}, which we shall now identify. If we differentiate Equation 7.9 with respect to W:

$$\frac{\partial E\left[|e|^2\right]}{\partial W} = 2|W|E\left[|x|^2\right] - 2E\left[nx^*\right] \qquad (7.10)$$

then the derivative will be zero at the minimum value of the mean square error. Equating Equation 7.9 to zero allows us to solve for the optimal filter W_{opt}

$$W_{opt} = \frac{E\left[nx^*\right]}{E\left[|x|^2\right]} \qquad (7.11)$$

The expected value of nx^* is proportional to the cross power spectral density function between n and x, and the expected value of the magnitude square of x is proportional to the (auto) power spectral density of x. (See Chapters 1 and 2). So Equation 7.11 may be rewritten as:

$$W_{opt} = \frac{S_{nx}^*(\omega)}{S_{xx}(\omega)} \tag{7.12}$$

which is (by definition) the transfer function between the reference signal x (interpreted as input) and the noise signal n (interpreted as output). In other words, the optimal configuration of the cancelling filter W_{opt}, is the inverse of the filter relating n (input) and x (output). This is not surprising, as the filter W here is used to obtain the noise signal, or an approximate of it, from the reference signal x.

The filter W_{opt} here is optimal in the sense of minimum mean square error (MMSE) and is derived from the statistics of signals taking into account the system property. This type of linear optimal filter achieved by MMSE is called the Viener filter, which is the most representative optimal filter, i.e. it is designed to minimise the mean square error for given signals and systems, which are wide sense stationary. The framework of Viener filters is optimal but not adaptive, i.e. Viener filters do not track noise and update the filter coefficient over time. In handling time varying noises and time variance of systems (may also be viewed as a kind of noise), it would be ideal if the filter could track the changes and update continuously or in short time intervals (in the case of discrete systems).

This can be seen in Figure 7.10, in which the transfer function e/s is seen by inspection to be trivially 1; the noise added at the first summing node is cancelled perfectly at the second summing node.

Such perfect performance is never achieved in practice for several reasons, of which the most important among them are:

1. Imperfect implementation of the cancelling filter.
2. Imperfect correlation between the noise and the reference.

We shall examine imperfect correlation in detail. Imperfect correlation between n and x can be modelled by the system depicted in Figure 7.11, in which the additional noise n_2 represents those components of x that are not correlated with n (we further

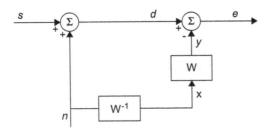

FIGURE 7.10 Operation of the noise canceller in the idealised case of perfect correlation between n and x

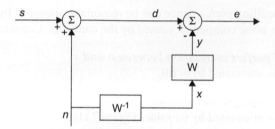

FIGURE 7.11 Noise cancelling with imperfect correlation between noise and reference. (Second uncorrelated noise signal, n_2, models uncorrelated component of x.)

assume that n_2 is independent of s). The LTI transfer function relating the correlated components of x to n is relabelled H for generality.

The error signal in Figure 7.11 is now:

$$E = s + n - W(Hn + n_2) \tag{7.13}$$

and the mean square error (which is still a positive definite quadratic function of W) is now:

$$E\left[|e|^2\right] = E\left[|s|^2\right] + E\left[|n|^2\right] + |W|^2 E\left[|Hn + n_2|^2\right] - 2\operatorname{Re}\left[WHE\left[|n|^2\right]\right] \tag{7.14}$$

where the uncorrelated terms that vanish have been set to zero. Differentiating Equation 7.14 and equating to zero yields the optimal cancelling filter for the configuration of Figure 7.11:

$$W_{opt} = \frac{H^* E\left[|n|^2\right]}{E\left[|H|^2 |n|^2 + |n_2|^2\right]} \tag{7.15}$$

Again, we shall introduce the auto and cross spectra. But in addition, we shall also bring in the ordinary coherence γ function that we used when dealing with transfer function measurement in noise:

$$\gamma_{n,x}^2 = \frac{|S_{nx}|^2}{S_{nn}S_{xx}} = \frac{|HS_{nn}|^2}{S_{nn}\left(|H|^2 S_{nn} + S_{n_2 n_2}\right)} \tag{7.16}$$

This, then, gives the optimum cancellation condition as:

$$W_{opt} = \frac{\gamma_{n,x}^2(\omega)}{H(\omega)} \tag{7.17}$$

Notice that the introduction of the uncorrelated noise does not change the phase of the optimal filter (defined by the inverse of H); it simply scales the optimal canceller gain by the coherence. (Remember that the coherence function is purely real).

The noise cancelling performance can be reasonably assessed by calculating the attenuation of the noise component caused by the canceller. Consider two cases:

In the case of perfect correlation between n and x
The noise is attenuated by ∞ dB.

In the case of imperfect correlation
The noise is attenuated by (consider Figure 7.11):

$$10\log_{10}\left[\frac{S_{nn}(\omega)}{S_{nn}(\omega)|1-WH|^2+|W|^2 S_{n_2 n_2}(\omega)}\right] \tag{7.18}$$

which, assuming the optimal filter W_{opt}, is used, can be written as:

$$10\log_{10}\left[\frac{S_{nn}(\omega)}{S_{nn}(\omega)\left(1-\gamma_{n,x}^2\right)^2+S_{n_2 n_2}(\omega)\gamma_{n,x}^2/|H|^2}\right]$$

$$= 10\log_{10}\left[\frac{1}{\left(1-\gamma_{n,x}^2\right)^2+\left(1-\gamma_{n,x}^2\right)\gamma_{n,x}^2/|H|^2}\right] \tag{7.19}$$

Equation 7.19 shows that the attenuation of the noise component on d is a function of the coherence between the noise and reference signal (and the magnitude of the transfer function H). The attenuation increases as the coherence increases (although, of course, the coherence function is bounded in magnitude between zero and 1.) The relationship between attenuation and coherence is plotted in Figure 7.12, in which the magnitude of H has been normalised to unity.

Note that useful levels of noise attenuation can only be achieved with high coherence between the reference and the noise signal to be cancelled—this poses a particular problem in electroacoustic applications. Given the expression in Equation 7.19 for the amount of noise attenuation expected from the cancelling system, the expected

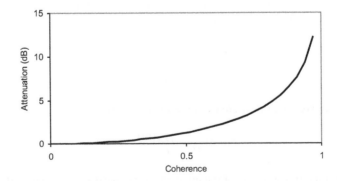

FIGURE 7.12   The attenuation possible for various degrees of coherence between noise and reference

improvement in speech-to-noise ratio is numerically equal to the amount of noise attenuation achieved (the speech being unaffected by the system when the noise and speech are truly uncorrelated).

7.3.2   A DISCRETE-TIME FORMULATION OF OPTIMAL FILTER

Implementing the optimal cancelling filter is a non-trivial task, and noise cancellation only really became feasible when electronic components allowed the filter to be implemented digitally. This not only allowed a precise, repeatable filter to be constructed but also allowed the system to become self-tuning, using adaptive filtering techniques (see next section). In order to understand the contemporary adaptive filtering approaches to noise cancelling, it is necessary to recast the (frequency domain) analysis of the previous section into a (discrete) time analysis—this is the purpose of the present section.

In discrete time form, the error sequence in Figure 7.9 is given by:

$$e_k = s_k + n_k - W_k^T X_k \tag{7.20}$$

in which k is the iteration index i.e. time, a lowercase variable is a scalar, and an uppercase variable is a vector. The superscripted T denotes transposition, such that $W^T X$ is the scalar product of the impulse response vector of the filter W and the vector of present and past reference inputs, X. The length of W and X is L, such that Equation 7.20 could be written as:

$$e_k = s_k + n_k - \sum_{j=0}^{L-1} w_j x_{k-j} \tag{7.21}$$

So the last term in Equation 7.20 is a normal filtering process, and this writing with transposition is essentially a matrix representation of a filtering process you should now be fairly familiar with. We write in this form because it is standard practise in adaptive filtering, not just to confuse! The squared instantaneous error is:

$$e_k^2 = s_k^2 + n_k^2 + 2s_k n_k - 2s_k W_k^T X_k - 2n_k W_k^T X_k + W_k^T X_k X_k^T W_k \tag{7.22}$$

If we make the same assumptions about the statistical relationship between the signals as in the previous section and apply averaging, the expected squared error is:

$$E\left[e_k^2\right] = E\left[s_k^2\right] + E\left[n_k^2\right] - 2W^T E\left[n_k X_k\right] + W^T E\left[X_k X_k^T\right] W \tag{7.23}$$

Notice that the expected squared error is a positive real quadratic function of the L dimensional space spanned by W—it has a minimum value at a unique position, W_{opt}, which we identify by the same process of differentiation and equating to zero as before. The derivative is:

$$\frac{\partial E\left[e_k^2\right]}{\partial W} = -2E\left[n_k X_k\right] + 2E\left[X_k X_k^T\right] W \tag{7.24}$$

which leads to an optimal impulse response vector for the transversal (FIR) filter W of:

$$W_{opt} = \frac{E[n_k X_k]}{E[X_k X_k^T]} \qquad (7.25)$$

in which $E[nX]$ is the $1*L$ cross covariance vector between n and x and $E[XX^T]$ is the $L*L$ (auto) covariance matrix of x. (Note the subscripts "k" disappear if the signals are stationary). Equation 7.25 is the discrete-time equivalent of Equation 7.12 (and has been written in similar notation to emphasise the similarity).

The computation of the coefficients of the optimal filter, defined by the impulse response vector W_{opt}, is a non-trivial computational task, as a result of:

- The averaging process (implied by the expectation operators) in the computation of the covariances;
- The matrix inversion.

Fortunately, a computationally efficient iterative approach to the identification of W_{opt} exists—this is studied in the next section.

7.3.3 ADAPTIVE METHODS AND LMS ALGORITHM

We have noted that the expected squared error associated with the noise cancelling system of Figure 7.9 is a simple quadratic function of W. Although it is possible to identify the optimal W, associated with the minimum expected squared error, in a one-step process (Equation 7.12), it is possible to find the minimum value using an iterative technique, using gradient searching methods.

The "surface" defined by the function in Equation 7.14 has a value of expected squared error associated with every filter W. If the current filter is W_j, then we may refine the filter design by moving in the direction of the negative gradient ("falling downhill" towards the minimum at W_{opt}):

$$W_{j+1} = W_j - \alpha \frac{\partial E[e^2]}{\partial W} \bigg|_{W=W_j} \qquad (7.26)$$

in which α is a positive scalar that determines the step size or update rate and the subscript j is an iteration index for the update process (the averaging processes invoked in Equation 7.26 may not allow the update to occur at each sample instant, i.e. $j \neq k$).

Since we have an expression for the gradient (Equation 7.24), we may substitute this into Equation 7.26 to give:

$$W_{j+1} = W_j + \alpha \left[2E[nX] - 2E[XX^T]W_j \right] \qquad (7.27)$$

which will converge towards W_{opt} as long as α is sufficiently small to make the process stable. The stability of the search process can (in this case) be examined analytically.

Stability is an important concern of adaptive algorithms. Rewriting Equation 7.27 as:

$$W_{j+1} = \left[[1] - 2\alpha E\left[XX^T \right] \right] W_j + 2\alpha E[nX] \qquad (7.28)$$

(where [1] is the $L*L$ identity matrix) shows that the search process for any of the L elements of W is described by a set of coupled difference equations (the coupling is achieved through the off-diagonal terms $E[XX^T]$). Stable bounds for the convergence of these equations can be assessed if the statistics are known. For example, if X is a white noise sequence, in which case:

$$E\left[XX^T \right] = \sigma^2 [1] \qquad (7.29)$$

(where σ^2 is the variance), then the equations un-couple to:

$$w_{i,j+1} = \left(1 - 2\alpha\sigma^2 \right) w_{i,j} + 2\alpha E\left[nx_i \right] \qquad (7.30)$$

(where $w_{i,j}$ is the ith element of Wj and $E[nx_i]$ is the i'th element of the cross covariance vector $E[nX]$). Equation 7.30 can be z-transformed to yield:

$$w_i(z) = \frac{2\alpha E[nx_i]}{z - \left(1 - 2\alpha\sigma_{xx}^2 \right)} \qquad (7.31)$$

which is stable when

$$0 < \alpha < \frac{1}{\sigma_{xx}^2} \qquad (7.32)$$

The gradient searching method of Equation 7.27 has the advantage of simplicity and assured convergence, but it is computationally expensive if either the system is changing or the signals are not stationary on the timescale of the adaptation of the filter, in which case the covariances have to be estimated continually. As both of these conditions are likely to be met in most real audio systems, an alternative approach may be more useful. Such an approach can be generated by attempting to minimise the instantaneous squared error (Equation 7.22) rather than the expected squared error.

Stochastic Gradient Search Methods

A search strategy that uses:

$$W_{k+1} = W_k - \alpha \left. \frac{\partial e_k^2}{\partial W} \right|_{W=W_k} \qquad (7.33)$$

has been found to converge to W_{opt} (provided the step size parameter α is suitably chosen). Substituting for the derivative gives:

$$W_{k+1} = W_k - \alpha\left[-2n_k X_k + 2X_k X_k^T W_k\right] = W_k - 2\alpha\left[(y_k - n_k)X_k\right] \qquad (7.34)$$

(Y is defined in Figure 7.9). Noting that $(y - n)$ is the error when $s = 0$ (and is equivalently the "correlated" error even in the presence of the non-zero but uncorrelated s) allows Equation 7.34 to be rewritten as:

$$\textbf{Key Equation: } W_{k+1} = W_k + 2\alpha e_k X_k \qquad (7.35)$$

This is the *least-mean-square (LMS) algorithm* discovered by Widrow and Hoff. It has been found to be robustly stable in many practical applications and is clearly a simple and computationally efficient approach to identifying W_{opt}. It forms the basis of most contemporary adaptive noise cancelling systems.

The performance of the LMS algorithm is illustrated below by example and also in the MATLAB® script *simple_lms.m*. Load MATLAB and run the script.

A simulation of a discrete time implementation of Figure 7.9 was coded, in which:

$$n_k = 0.5x_k + 0.2x_{k-1} \qquad (7.36)$$

and an $L = 2$ adaptive filter W was updated using the LMS algorithm (Equation 7.35) in a signal environment in which s was a simple sinusoid and x a random process. The error signal is shown in MATLAB Figure 3.

The initial noise is seen to be quickly cancelled, leaving a pure sinusoid—the signal s. Notice that the decay of the noise follows a roughly exponential form. This is due to the fact that the convergence behaviour of the LMS algorithm approximates the first order convergence of the true "steepest descent" algorithm. However, precise stable bounds for the LMS algorithm can only be determined in the context of certain simple deterministic reference signals.

The convergence of the two elements of the $L = 2$ impulse response vector (the weight of the adaptive filter W) are shown in MATLAB Figure 4. The weights are seen to approach the optimal values implied by Equation 7.36.

This is very similar to the system shown in Figure 7.10, except that the white noise signal is fed direct to x and then filtered to get the signal n. The script *simple_lms2* actually does it the way around shown in Figure 2; it takes a little time to run. This also shows how the noise remaining in the output reduces over time in a mean square plot.

Having studied fundamentals of noise cancelling, we now consider the practicalities of a number of applications within speech communication systems.

Noise Cancelling on the Microphone Signal

Consider the situation when additive noise is detected by the microphone, as depicted by Figure 7.13.

The speech signal at the microphone is acoustically summed with an additive noise n generated by the noise source. A reference signal x is obtained from the noise source and passed to a cancelling filter that produces the cancelling signal y.

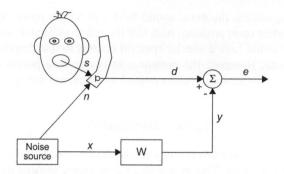

FIGURE 7.13   Noise cancelling on a telephone microphone

Given the similarity between Figures 7.9 and 7.13, it is clear that an LMS algorithm would adjust W to the optimal configuration to minimise the noise component on the "error" signal e, leaving speech if the signals in Figure 7.13 satisfied the requirements we have already identified:

- The speech and noise must be uncorrelated.
- The speech and reference must be uncorrelated.
- The noise and reference must be correlated and related to each other through a transfer function $H\,(=x/n)$ that has a causally stable inverse (since the optimal configuration of W is H^{-1}).

Satisfying the second requirement above is potentially difficult as the output of a general noise source is an acoustic signal that will need to be transduced by a microphone. Unfortunately, such a microphone would have some sensitivity to the speech signal (which would generate a component of n correlated with s and so offend the second requirement above). This suggests that the reference signal should be transduced by a microphone with minimum sensitivity to the speech (possibly a directional type oriented with a response "null" towards the speaker) positioned as far as possible from the speaker.

There are some situations where the reference signal can be detected by other than acoustic means (such as the use of an electronic tachometer signal to derive a reference for engine rotational order-related noise in a car application), in which case there will be no speech corruption of the reference signal.

Positioning a reference sensing microphone far from the speaker (as suggested by the requirement to minimise speech contamination of the reference, above) carries some practical penalties.

In a small enclosed sound field, such as in a room or a car, the low frequency sound field is strongly spatially dependent as a consequence of the standing wave patterns associated with the normal modes of the space. This means that the microphone used to detect the reference signal may receive a radically different noise power spectrum than the noise power spectrum detected by the microphone within the telephone handset. This would need to be corrected by the cancelling filter W and would practically degrade cancelling performance.

At higher frequencies, the noise sound field will have a power spectral density that is less dependent upon position, such that the reference signal detected by a second microphone would have a similar spectral content to that detected by the main speech microphone. However, the pressures at two remote points in an enclosed sound field become increasingly uncorrelated as frequency (or spacing) increases. In fact, in the limiting case of a "diffuse field," the coherence function falls off as:

$$\gamma_{nx}^2(\omega) = SINC^2(\omega d/c) \tag{7.37}$$

where d is the spacing between the two microphones, *sinc* is the $sin(x)/x$ function, and c is the speed of sound. This means that the reference sensing microphone must be close to the main microphone to ensure good correlation between n and x. It is seen that satisfying the second and third statistical requirements above suggest conflicting microphone positions for the reference sensing microphone; clearly, the performance of the system of Figure 7.13 is practically limited by this conflict!

Noise Reduction at the Ear

When a telephone is used in a noisy situation, the sound at the ear is due both to the telephone loudspeaker and the background noise. Given that an electrical signal is available that states what sound should be at the ear (the clean telephone signal alone), it is possible to arrange for the loudspeaker of the telephone to cancel some component of the pressure due to the background noise around the earpiece (and hence, when in use, at the user's ear). This is illustrated in principle in Figure 7.14.

A miniature microphone, having electroacoustic response M (Volts/Pascal), is placed on the body of the telephone at the position of the ear in normal use. The response of this microphone is:

$$x = M(\omega)[n + L(\omega)T(\omega)d] \tag{7.38}$$

in which n is the background noise pressure at the microphone, d is the electrical drive to the loudspeaker, L is the effective loudspeaker electroacoustic response $(m^3V^{-1}s)$, and T is the acoustic transfer impedance between the loudspeaker and the

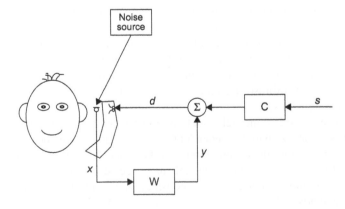

FIGURE 7.14 Active noise reduction at the earpiece of a telephone

microphone (Pascal, seconds/cubic metre). L, T, and M are shown explicitly as functions of frequency in Equation 7.38 as the precise details of the frequency response of the loop LTM will become important to the operation of the noise reduction strategy.

The microphone signal x is used to design a component of the loudspeaker drive signal d that is intended to minimise the noise pressure detected by the microphone. Specifically:

$$d = Wx + Cs$$

$$\Rightarrow x = Mn + MLTWx + MLTCs$$

$$\Rightarrow x = \frac{Mn + MLTCs}{1 - MLTW} \tag{7.39}$$

$$= \frac{Mn}{1 - MLTW} + \frac{MLTCs}{1 - MLTW}$$

Equation 7.39 shows that x (and, therefore, the noise at the user's ear) has two components: a component associated with the noise and a component that is related to the signal s. If we assume that the noise and s are uncorrelated, we can deal with the components individually. When the filter W of Figure 7.14 is turned off ($W = 0$), then the noise component of x is Mn. When W has a non-zero response, the noise component is:

$$x_{noise} = \frac{Mn}{1 - MLTW} \tag{7.40}$$

The action of the filter W is, therefore, to change the noise power spectrum by:

$$attenuation = 10 \log_{10} \left[|1 - MLTW|^2 \right] \tag{7.41}$$

In other words, if W is designed correctly, the action of W is to attenuate the noise pressure caused by the background noise source at the location of the microphone. Although non-zero W also modifies the speech component of x relative to the $W = 0$ case, the filter C can be designed to compensate for this modification, equalising the speech response.

Equation 7.41 suggests that the attenuation is maximised when the loop gain $MLTW$ is maximised.

This is the case although the loop must be stable for useful operation; that is to say, W must be designed with reference to the available electroacoustics, MLT, such that:

$$|1 - MLTW(\omega)| < 1 \; when \; \angle(1 - MLTW(\omega)) = 0 \tag{7.42}$$

This requirement for loop stability limits practically the noise reduction possible using this technique, although measurements on a mobile telephone handset suggest that some useful attenuation is possible. The performance of the noise reduction system of Figure 7.14 can be further improved if signal processing techniques are applied to effectively "break the feedback loop, in which case the adaptive methods described previously may be applied."

7.4   GENETIC ALGORITHMS: ANOTHER ADAPTIVE TECHNIQUE

In the last section, you were introduced to the concept of a signal processing algorithm adapting to the signals it receives. The LMS algorithm was searching to find the optimal filter that minimises the noise in a signal. This is an example of numerical optimisation, an algorithm where a computer searches to find a combination of numbers (in this case, filter taps) to minimise a figure of merit (in this case, the amount of noise in a system).

While it is possible to use numerical optimisation in active noise control, it is most useful in design work. For example, consider the system below:

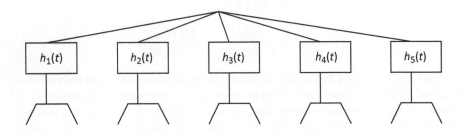

A signal is fed to five loudspeakers via a bank of five digital filters. By altering the digital filter coefficients, we can change the radiation characteristics of the array. What would be the most appropriate filter coefficients A to achieve omni-directional radiation? This is the type of task that we can ask a numerical optimiser to solve. But before reaching for such a tool, you need to consider:

- Is there an analytical solution to the problem? (In this case there is for an omni-directional characteristic.)
- What physical understanding of the problem do we have, and how can we capitalise on this to maximise the success of the algorithm? (Omni-directional array from highly directional loudspeakers is going to be difficult.)
- Is there a theoretical (analytical or empirical) model available that correctly models the real life situation?
- Will the theoretical model run fast enough?

Essentially, the problem is to find a set of numbers A (filter coefficients) that minimises some cost function ε (figure of merit, error parameter). In the case above, the cost function would measure the closeness of the radiated polar response to the desired radiated polar response. This could be a mean square parameter:

$$\varepsilon(A) = \sum_{m=1}^{M} \left(L_{d,m} - L_{a,m}\right)^2 \tag{7.43}$$

where there are M measurements on the polar response, $L_{d,m}$ is the desired levels in the polar response, and $L_{a,m}$ are the actual levels for the particular set of parameters **A**. Part of using an optimisation problem is determining the most appropriate cost function.

There are a number of different optimisers that can be used:

- If the gradient of the cost function is available, an optimiser that exploits the gradient should be used because that will be so much faster.
- Otherwise, if calculating the cost function is very fast ($< 1s$) and the number of parameters **A** small, it probably does not matter which optimisation algorithm one chooses.
- We will use a genetic algorithm (GA) because they are currently very popular and also rather fun!

7.4.1 Genetic Algorithms

A genetic algorithm mimics the process of evolution that occurs in biology. A population of individuals is randomly formed. Each individual is determined by their genes; in this case, the genes are simply the set of numbers **A** that describe the filter coefficients. Each individual (or set of coefficients) has a fitness value (figure of merit) that indicates how good they are at radiating in an omni-directional manner. Over time, new populations are produced by combining (breeding) previous filter coefficients, and the old population dies off. Offspring are produced by pairs of parents breeding, and the offspring has genes that are a composite of the genes from the parents. The offspring shape will then have features drawn from the parent shapes, in the same way that facial features of a child can often be seen in the parents. A common method for mixing the genes is called multi-point crossover. In the binary scheme described below, there is a 50% chance of the child's gene coming from parent A, and a 50% chance of the gene being from parent B.

If all that happened was a combination of the parent genes, then the system never looks outside the parent population for better solutions. The fish array would never get lungs and walk about on the land. As with biological populations, to enable dramatic changes in the population of array filters, mutation is needed. This is a random procedure whereby there is a small probability of any gene in the child sequence being randomly changed, rather than coming from the parents directly.

Selecting shapes to die off can be done randomly, with the least fit (the poorest arrays) being most likely to be selected. In biological evolution, the fittest are most likely to breed and pass on their genes, and the least fit are the most likely to die; this is also true in an artificial genetic algorithm. By these principles, the fitness of successive populations should improve. This process is continued until the population becomes sufficiently fit so that the array can be classified as optimum. A common termination criterion is when all the members of the population are identical.

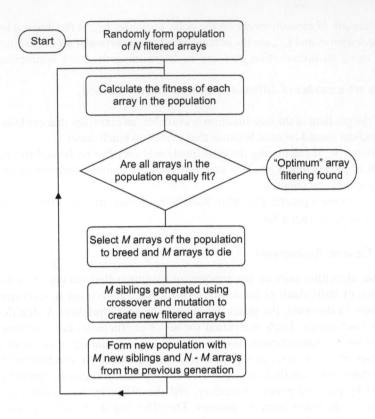

The process of a genetic algorithm is shown in the flow chart above. The MATLAB script *GA.m* demonstrates the use of a genetic algorithm for the example of an array of 13 point sources. There are many variations on a genetic algorithm in terms of how breeding, mutation, etc, are implemented. The script gives a fairly standard technique. It may not be the most efficient coding, but it demonstrates some important points about genetic algorithms.

The task is to make an omni-direction radiation from a set of 13 point sources. These point sources can have phases of +1 or −1. Consequently, each individual has 13 genes that are either +1 or −1, indicating the phases of the loudspeaker drivers. The fitness of a particular array is calculated via the standard deviations of the radiated energies over a polar response. If all receivers on the polar response received the same energy, then the fitness would be zero. Any deviations from the omni-directional response causes the standard deviation to increase.

The algorithm needs tuning in terms of mutation rate, population size, rate of dying, etc. This is annoying if you are using the algorithm for a one off-test—not so bad for repeat runs. In the example given, the population consists of 50 arrays. One-sixth of the population dies in each generation, and the fittest in the generation always survive. For convenience, the number breeding and the number dying are set equal. The choice of the particular individuals to breed and die is done via the cumulative probability distribution. Consider choosing an individual to die. For the nth individual, the cumulative probability distribution c_n is given by:

$$c_n = \frac{\sum_{m=n}^{N} f_m - \min(\mathbf{f})}{\max(\mathbf{f}) - \min(\mathbf{f})} \qquad (7.44)$$

where $\mathbf{f}$ is the matrix of fitnesses, with f_n being the fitness of the nth array (individual). There are assumed to be N individuals in the population. A dice is then rolled from 0 to 1 to determine which individual should die. As individuals with poor fitness, in this case large standard deviations, will be associated with large steps in the cumulative distribution, they are most likely to be chosen. The fittest individual, with the smallest fitness value, cannot be chosen in this scheme. There are other ways of choosing who should breed and die, but this method is fast and efficient.

When breeding, for each gene a dice is rolled from 0 to 1. If the dice value is > 0.5, then the gene comes from parent A; if the dice value is < 0.5, then the gene comes from parent B. The mutation rate is set so there is a 2% chance for each gene mutating. Whether a gene mutates is determined by rolling a dice, and if the gene mutates, the state is flipped from -1 to $+1$ or vice versa. This is quite a high mutation rate, but is fine because there are a small number of genes in this example.

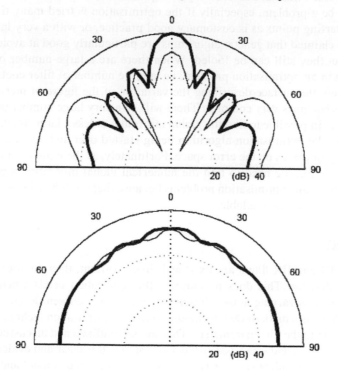

The radiation from four arrays. Top: Plane surface (bold) and linear array (normal). Bottom: GA solution (bold) and Barker array (normal).

The figure shows the polar responses from various arrays. It compares a line array, where all the point sources radiate in phase with the optimised solution produced by the genetic algorithm (GA). It shows that the GA produces more arrays that radiate more evenly in all directions. Also shown is an array when the phases are set completely randomly. Most interesting is the line labelled Barker. This is using an $N = 13$ length barker sequence (Fan & Darnell, 1996) to determine the filter coefficients. This is a number sequence with good autocorrelation properties, i.e. one whose Fourier transform is the flattest possible for a binary sequence. The significance here is that it would have been possible to guess this solution (labelled Barker) and get the coefficients from a text with a lot less effort than involved encoding a GA. If a possible good solution is known, it is a good idea to include one of these individuals in the starting population.

In any optimisation problem, there will be a large number of local minima, but somewhere there will be the numerically lowest point, called the global minimum. A common analogy for a two-dimensional optimisation is finding the lowest point on a hilly landscape (while blindfolded). The blindfolded person might find the nearest valley bottom and think the best point has been found, not realising that over the next mountain ridge there is a lower valley. The key to a good optimisation algorithm is not to be trapped in poor local minima, but to continue to find deep local minima. Provided a good optimisation algorithm is chosen, this should not be a problem, especially if the optimisation is tried many times from different starting points as is customary good practice, or with a very large population. It is claimed that genetic algorithms are particularly good at avoiding local minima, but they still can be fooled. When there are a large number of degrees of freedom in an optimisation problem, i.e. a large number of filter coefficients to be optimised, the surface describing the variation of the figure of merit with the coefficients becomes very complex. There will be a very large number of minima. It is virtually impossible to find the global minimum unless a large amount of time is used with the optimisation algorithm being started over and over again from a wide variety of places on the error space. Fortunately, as the number of degrees of freedom increase, the need to find the numerical global minimum becomes less important for many optimisation problems because there will be a large number of equally good solutions available.

SUMMARY

Classical FIR and IIR filters are fixed and time-invariant; they do not track environmental changes. This does not address the application needs when handling situations such as varying noises. Based on some wisely chosen design objectives, adaptive filters as a modern DSP technique employs optimisation techniques to enable adjustment to the filter parameters. This chapter outlined and discussed the LMS adaptive algorithm and some associated applications. It is clear and evident that key in the design of an adaptive filter is to derive or arrange the "input" and "desired" signals to fit in the adaptive filter scheme based on the actual applications. Adaptive filters make use of optimisation to achieve design objectives, typically minimised errors. The optimisation algorithm is often running on the fly to track the changes of environment, however in some other application, the optimisation can be a one-off process, e.g. when an adaptive filter is used to identify an unknown system. This chapter focused on the LMS filters. These represent the most popular adaptive filters found in many diverse applications. Other optimisation algorithms can be used for adaptive filters. Genetic algorithms are used as a fairly unusual example to highlight that once the error function is suitably derived, adaptive filters can be achieved using different optimisation techniques.

REFERENCE

Fan, P. and Darnell, M. (1996) *Sequence Design for Communication Applications*, John Wiley and Sons Inc.

BIBLIOGRAPHY AND EXTENDED READING

Hayes, M. H. (2008) *Statistical Digital Signal Processing and Modelling*, Wiley.
Haykin, S. and Widrow, B. (2003) *Least-Mean-Square Filters (Adaptive and Cognitive Dynamic Systems: Signal Processing, Learning, Communications and Control)*, Wiley.
Manolakis, D. G., Ingle, V. K., and Kogon S. M. (2005) *Statistical and Adaptive Signal Processing: Spectral Estimation, Signal Modeling, Adaptive Filtering and Array Processing*, Artech House.

EXPLORATION

1. *adaptive.zip* contains MATLAB codes for exploration. Unpack it in the working directory of MATLAB. Experiment with the simple LMS discussed in this chapter and run it. The update rate in an LMS algorithm is crucial to the system performance.
   - Try changing the value of alpha and examine the performance, in particular: What happens when alpha becomes too large and why? What happens if alpha is too small and why?
   - Try changing the amplitude of the interfering noise. Does this affect the ability of the LMS routine to achieve noise cancellation?
   - Try changing the input signal to noise. Does the LMS filter still work? What provides evidence for it working (the error doesn't!) or why does it not work?
   - Make the noise signal dependent, (e.g. $x = s$, as the simplest case). Does the LMS filter still work? Explain the findings.
2. Load up the script for the GA optimisation given in zipped file. There are many parameters that can change the performance of the algorithm:
   - Try changing the mutation rate (mutate_rate). Can you see evidence of mutation in the error verses iteration graph (Figure 1)? Does the system learn better with a smaller or larger mutation rate? (This effect of mutation is more obvious with the case considered in the assignment).

- What effect does the population size (npop) have on the learning? Remember the learning time isn't just determined by the number of iterations, because the bigger the population, the slower each iteration takes. You can time things in MATLAB using tic toc.
- If you run the algorithm several times, you get different solutions. Why is this?
- If you run the algorithm several times, but use the solution from the previous run as a starting member of the population for the next run, you will often get better solutions. Why? This may be a useful trick for the assignment.

MINI PROJECT IDEA

Consider the use of an adaptive filter to identify an unknown system. Genetic algorithm is used for optimisation in filter adaptation. To evaluate how the adaptive filter approximates an unknown system, it is suggested that an FIR filter, say a 21-tap FIR low-pass filter is designed first, and then used as "the system in question." As the system in question is assumed "unknown," there is not prior knowledge of its complexity. It is, therefore, necessary to empirically determine the number of taps of the adaptive filter to achieve a sufficiently accurate model of the "unknown" system.

8 Machine Learning in Acoustic DSP

Acoustic and audio signals are information rich. We often make decisions, take actions, and retrieve semantic meanings or even emotional cues from these signals. On hearing an alarm, we quickly take actions depending upon the type of alarm. We receive information and communicate with speech signals and enjoy music for its rich tonal characteristics and emotional cues that go way beyond what can be described using languages.

Since many decisions and judgments are made based upon acoustic signals, acoustic and audio DSP has, naturally, been extended to automate these processes. Instead of simplistic manipulations of signals, information of interest and criteria for decision making are extracted from the signals. Automated speaker recognition, i.e. a system that determines who the talker is (or more precisely, who the talker is likely to be), automated speech recognition, machine audition, and music information retrieval have now all become reality. There are also many bespoke acoustic or audio pattern recognition systems for various applications from scientific experiments to everyday uses. For example, acoustic signals can be used to localise sound sources and classify them. This has been used to monitor bird migration and behaviours by environmental and biological scientists. Acoustics signal analysis, information extraction, and pattern recognition have seen many more applications and are the key techniques used for underwater communications, sensing and imaging, and medical diagnosis and imaging. Acoustic pattern recognition is a universal term used to describe the use of computer algorithms to identify specific futures for decision making and information mining. Many pattern recognition algorithms are built upon so-called machine learning.

8.1 GENERAL CONCEPT OF ACOUSTIC PATTERN RECOGNITION

A general acoustic pattern recognition system involves signal acquisition, pre-conditioning, feature extraction, feature selection, and decision making. An acoustic signal of interest is picked up or monitored using a suitable transducer, e.g. a microphone, and pre-conditioned as necessary. The pre-conditioning may involve digitisation, de-noising, signal separation, pre-emphasis, band limiting, etc., for subsequent processing. The feature extraction derives new features from raw data with an intention to reduce redundancy and make the representation more informative; feature selection is typically an algorithm that chooses a subset of features that can be used to achieve the design goals of recognizing the pattern of interest with certain criteria and to facilitate decision making. For example, if a system is designed to recognise speech signals from other audio content, zero crossing rates (which will be discussed later in this chapter) might be used. In the general field of machine

FIGURE 8.1   A block diagram of a general acoustic signal pattern recognition system

learning and pattern recognition, feature extraction and feature selection are two different stages, serving distinctive purposes and using different algorithms. Feature selection often means an algorithm that automatically chooses subsets of features with certain metrics. It is interesting to note that in acoustics, especially audio pattern recognition, feature selection algorithms are not particularly popular in use. Over the past few decades, a good number of common audio and acoustic features have been established and tested by a good number of researchers; these features typically have either clear physical or perceptual meanings. Researchers in the field of audio and acoustics tend to adopt features with known meanings heuristically and manually; as a result, in Figure 8.1, the feature extraction and selection are combined in one block. The final decision-making stage is typically a classifier, which differentiates the acoustic pattern of interest from others. Decision making can be as simple as a threshold, but in most cases, due to the complexity and stochastic nature of acoustic signals in the real world, statistical machine learning is a popular choice.

8.2   COMMON ACOUSTIC FEATURES

Features are often chosen heuristically as a starting point; since most signals we handle are random processes, there is no analytic mathematical model available for them. Empirical or statistical approaches are adopted to "validate" these heuristically chosen features. Over decades of acoustic pattern recognition research and development, a good number of features have been calculated and tested for various tasks. It is impossible and also unnecessary to exhaustively list them. Some of the most representative ones are highlighted and discussed in this section. Features may be categorised according to the domains in which they are represented. Commonly used features can be in the time, the frequency, the time-frequency, or other domains. Certain characteristics of signals may be more apparent in one particular domain than in the others; this is one of several reasons of manipulating signals in the transformed domains. There are also psychoacoustically shaped audio signal representations, which intend to better relate themselves to human perception of sounds.

8.2.1   ACOUSTIC FEATURES AND FEATURE SPACES

Features refer to statistical characteristics of signals. Features of acoustic and audio signals may or may not be directly related to perception. A feature vector is an n-dimensional vector, either a row or column vector, of numerical features that describe the signals under investigation. The vector space associated with these vectors is said to be the feature space. It is evident that features are characteristics extracted statistically from the signals. Feature extraction and selection function reduce data points or dimensionality, but retain or highlight the information of interest. A good feature space makes hidden information revealing and reduces the burden on subsequent decision-making mechanisms. Signals can be represented in the time, frequency, or time-frequency domains. Accordingly, the features can be in one of these domains.

Many signal statistics in the time and frequency domains as discussed in Chapter 1 and in Chapter 2 can be, and are, in fact, commonly used as features. Amongst them, the mean, variance and moments, DFT coefficients, power spectra, probability density functions are the popular and significant ones to adopt. In addition to these, there are many other well-established features.

8.2.1.1   Time-Domain Features

For acoustic and audio signals, the time domain is the native domain in which signals are acquired, sampled, and digitised. In a time-domain representation of signals, the abscissa is the time and the ordinate is the amplitude. As the native domain of signals, sometimes features in the time domain are expressive and transparent.

Since audio signals are not stationary generally, short analytical frames of, say 20-40 ms, are used so that signals within the frames can be viewed as stationary and statistical features can be acquired. While 20 to 40-ms short windows are common-place for detailed analysis of audio signals, longer-term statistical features are also used occasionally. It is, therefore, important to specify the duration (window width) within which the features are acquired. The root mean square and crest factor (to be discussed later in this subsection) are good examples for the need to differentiate short-term and long-term statistics of signals.

In the following definitions, $s(n)$ denotes the nth sample of a signal, N is the total number of samples, and Fs is the sampling frequency.

1. **Zero Crossing Rate**

   Zero crossing rate (ZCR) is a basic property of an audio signal. Zero crossing of a signal is found by calculating the times that it crosses the time axes, i.e. sign changes. When the zero crossing counts are normalised to the length of the signal (number of samples or time duration), the zero crossing rate zc is obtained by

   $$zc = \frac{1}{2}\left(\sum_{n=0}^{N-1}\left|sign(s(n)) - sign(s(n-1))\right|\right)\frac{Fs}{N} \qquad (8.1)$$

   where the $sign(x)$ function is defined as:

   $$sign(x) = \begin{cases} 1 & if \quad x > 0 \\ 0 & if \quad x = 0 \\ -1 & if \quad x < 0 \end{cases} \qquad (8.2)$$

   ZCR has been an established feature for speech signals for decades. It is often used as a feature to discriminate speech and music; for speech, it can be further used to detect voiced and unvoiced segments. ZRC is typically counted in short analytical frames of 20 to 40 ms. In particular, the short-term variance of the ZCR was found even more effective than ZCR itself in certain tasks (Gerhard, 2003).

2. **Root Mean Square Value**

   Root mean square (RMS) has been used to determine the DC equivalent of varying signals in terms of energy. Both short-term and longer-term RMS values of signals are used to characterise the signals; therefore, the duration RMSs measure should be carefully specified to avoid ambiguity. For a time discrete signal $s(n)$, the RMS value S is given by

$$S = \sqrt{\frac{1}{N} \sum_{n=0}^{N-1} s(n)^2} \tag{8.3}$$

RMS levels (when RMS values are expressed in decibels) of a signal are directly related to the sound pressure levels. In the literature, some authors refer to the RMS level as "loudness." However, in a strict sense, loudness is a measure of sound perception, not a physical measure of signals themselves. The relationships between sound pressure level and perceived loudness have been established; for example, the equal loudness contours give the relationships between the SPL (dB) and loudness (phon) for pure tones. The "loudness" used by some authors to mean RMS should not be confused with psychoacoustically defined loudness.

3. **Crest Factor**

   Crest factor cf, introduced in Chapter 3 when we discussed the stimulus signals for measurements, is commonly used to describe the dynamics of a signal. For deterministic signals that can be described by a mathematical function, it is straightforward. For a random signal, the crest factor is determined from the envelope of the signal. In short, the crest factor is defined as the peak-to-RMS ratio.

$$cf = 20 \log \frac{S_{peak}}{S_{rms}} (\text{dB}) \tag{8.4}$$

where S_{peak} is the peak value (i.e. the local maxima) found in the signal envelope obtained by a short-time RMS detector or other envelope detectors, e.g. the envelope calculated from a Hilbert transform algorithm, and S_{rms} is the RMS value of the signal calculated over a longer (significantly longer than the short-time RMS detector) time. These should not be confused with the maximum value found from individual sample $s(n)$. Depending upon the application, one may also use the average of peak RMS values found in the signal excerpt, i.e.

$$cf = 20 \log \frac{\overline{S}_{peak}}{S_{rms}} (\text{dB}) \tag{8.5}$$

4. Entropy

Entropy was first introduced by Shannon to determine the amount of information included in an excerpt of a signal, so it is also known as "Shannon's entropy" to differentiate it from the entropy used in thermodynamics. It was developed in the context of information theory for telecommunications. In a simplistic way, the entropy H can be interpreted as the minimum number of bits that are needed to completely encode the information carried in a signal. In the case of binary encoding, 2 based log is used:

$$H(s) = H(p) = \sum_{n=0}^{N-1} p(s) \log_2 \frac{1}{p(s)} \tag{8.6}$$

where p is the probability distribution of signal s. To adopt notations in common use, $H(s)$ indicates that entropy H is a function of $s(n)$; this *should not be confused with Laplace transfer function.*

Entropy can be used to indicate the amount of information included in an excerpt of a signal and sometimes the number of sources or complexity of the signal.

Note that if we have two random variables s_1 and s_2, the joint entropy between them becomes

$$H(s_1, s_2) = \sum_{n=0}^{N-1} p(s_1, s_2) \log_2 \frac{1}{p(s_1, s_2)} \tag{8.7}$$

If s_1 and s_2 are independent, then $H(s_1,s_2) = H(s_1) + H(s_2)$. This implies that entropy can be used to indicate the number of independent sources.

Strictly speaking, the entropy is not a time-domain feature, since it is a not a function of time or samples. It is not a frequency-domain feature either, as there is no variable indicating frequencies. However, the entropy is often determined frame by frame, or block by block along the time, resulting in a sequence that represents the entropies over time. Many researchers loosely classify the entropy as a time-domain feature.

8.2.1.2 Frequency-Domain Features

Frequency-domain features are often related to the perception of pitch, bass, treble and warmth of sounds. They are mostly derived from Fourier-transformed domains, but many other transforms are possible and found useful in some cases.

For stochastic signals, statistical features are established by the analysis of signal segments in short windows so that essential stationarity can be assumed. Window widths of 20 to 40 ms are typical for speech, music, and other audible sounds. Hamming and Hanning windows with 50% are popular choices. For other acoustic signals, different window widths may be adopted.

The Hamming window

$$w(n) = 0.54 - 0.64 \cos\left(\frac{2\pi n}{N-1}\right), \qquad n = 0,1,...,N-1 \qquad (8.8)$$

is used as an example here, in which n is the sample index of the window and N is the length of the window. The window $w(n)$ is moved along and multiplied by the signal $s(n)$ to obtain the windowed segment $x(n)$. Fourier transform is performed:

$$X(k) = \sum_{n=0}^{N-1} x(n)e^{-j2\pi k \frac{n}{N}}, \qquad k = 0,1,...,N-1 \qquad (8.9)$$

where $X(k)$ is the Fourier transform of $x(n)$ and k is the index of frequency bins.

A number of frequency-domain features can be further extracted from the $X(k)$. Examples are examined next.

1. **Spectral Roll-off Point**

   The spectral roll-off (SR) point is a specific frequency corresponding to the frequency bin K_{roll} that 85% of the total spectrum magnitudes are concentrated below K_{roll}, i.e.

$$\sum_{k=0}^{K_{roll}} |S(k)| = 0.85 \sum_{k=0}^{N_{FT}/2} |S(k)| \qquad (8.10)$$

   where N_{FT} is the number of points for the DFT. Here N_{FT} and N may or may not be identical since zero padding may be used.

2. **Spectral Flux**

   The spectral flux (SF) is the value of the average variation in signal spectrum between adjacent frames; it measures the local spectral change. It is defined as

$$SF = \frac{1}{L \cdot N_{FT}} \sum_{k=0}^{L-1} \sum_{k=0}^{N_{FT}^{-1}} \left[log\left(|S_l(k)| + \delta\right) - log\left(|S_{l-1}(k)| + \delta\right) \right]^2 \qquad (8.11)$$

   where $S_l(k)$ is the DFT of the lth frame, N_{FT} is DFT order, L is the total number of signal frames, and δ is a small parameter to avoid overflow.

   Speech SF values are higher than music ones and the environment sound has the highest value. Also, the environmental sound changes dramatically between successive frames (Lu et al., 2002). The SF has been used for speech/music discrimination applications e.g. (Burred & Lerch 2003; Scheirer & Slaney, 1997).

3. **Spectral Centroid**

The spectral centroid (SC) is used to characterise the spectrum and it has no relation to the signal harmonic structure. It is calculated as the weighted average of the discrete frequencies present in the signal, determined by DFT.

$$SC = \frac{\sum_{K=0}^{NFT/2} f(k) P_s(Kk)}{\sum_{K=0}^{NFT/2} P_s(K)} \tag{8.12}$$

where P_s is the power spectrum for the segment, $f(k)$ is the frequency of the kth bin, and N_{FT} is the size of the DFT. It is designed to discriminate between different musical instrument timbres (Kim et al., 2005).

4. **Audio Spectrum Centroid**

The audio spectrum centroid (ASC) measures the centre of gravity of a log-frequency power spectrum. The summation of all power coefficients below 62.5 Hz will be calculated in order to prevent having disproportionate weight for DC components and the very low-frequency components.

In the discrete frequency bins scale, the values are below the following index:

$$K_{low} = floor(62.5/\Delta F) \tag{8.13}$$

where $floor(x)$ is a function that returns the largest integer that is less than or equal to x, and $\Delta F = F_s/N_{FT}$ represents the frequency interval between two consecutive FFT bins. The resulting power spectrum $P'(k')$ is related to the original spectrum $P(k)$, and the relation is given by

$$P'(K') = \begin{cases} \sum_{k=0}^{K_{low}} P(k) & \text{for } k' = 0 \\ P(k' + K_{low}) & \text{for } 1 \leq k' \leq \dfrac{N_{FT}}{2} - K_{low} \end{cases} \tag{8.14}$$

The frequencies $f'(k')$ corresponding to the new bins k' are given by

$$f'(K') = \begin{cases} 31.25 & \text{for } k' = 0 \\ f(k' + K_{low}) & \text{for } 1 \leq k' \leq \dfrac{N_{FT}}{2} - k_{low} \end{cases} \tag{8.15}$$

where $f(k)$ is the discrete frequency corresponding to bin indexes k and k is the nominal frequency of the low frequency coefficient. It has been selected at the centre of the low-frequency band $f'(0) = 31.25$ Hz.

Now the ASC can be defined by the following relation between the modified power coefficients $P(k)$ and their corresponding frequencies $f(k)$ as

$$ASC = \frac{\sum_{k'=0}^{(N_{FT}/2)-K_{low}} log_2\left(\frac{f'(k')}{1000}\right)P'(k')}{\sum_{K'=0}^{(N_{FT}/2)-K_{low}} P'(K')} \tag{8.16}$$

Each modified power spectrum coefficient $f'(k')$ is weighted by the corresponding power coefficient $P'(k')$.

The ASC provides information on the power spectrum shape. It indicates whether a power spectrum is dominated by high or low frequencies; also, it gives an approximation of the signal perceptual sharpness. Finally, the log-frequency scaling utilised in the ASC definition functions to approximate the frequency-dependent characteristics of human auditory system (HAS)(Kim et al., 2005).

5. **Audio Spectrum Spread**

The audio spectral spread (AS)S measures the spectral shape; it has been defined in the MPEG-7 standard as the second central moment of the log-frequency spectrum. ASS can be found by taking the RMS deviation of the spectrum from its centroid ASC:

$$ASS = \sqrt{\frac{\sum_{k'=0}^{(N_{FT}/2)-K_{low}}\left[log_2\left(\frac{f'(k')}{1000}\right)-ASC\right]^2 P'(k')}{\sum_{k'=0}^{(N_{FT}/2)-K_{low}} P'(k')}} \tag{8.17}$$

where $P'(k')$ is the modified power spectrum coefficients (Equation 8.14) and $f'(k')$ is the corresponding frequencies. (Equation 8.15)

The ASS indicates the distribution of the spectrum around its centroid, so a low ASS value means the spectrum power might be concentrated around the centroid, while a high value means the spectrum power might distributed across a wider range of frequencies. It is designed specifically to help differentiate between noise-like and tonal sounds (Kim et al., 2005).

6. **Spectral Entropy**

Spectral entropy returns the relative Shannon entropy of the spectrum. The Shannon entropy, used in information theory, is based on the following equation

$$H(P) = -\left(\frac{sum(p \cdot \log(p))}{\log(length(p))}\right) \tag{8.18}$$

Many other signal statistical measures reviewed in Chapter 1, such as expected value and moments, can all be used as features. We do not need to repeat them here.

7. **Pitch**

There are many techniques to approximate the pitch; one way is to take a series of short-time Fourier spectrum frames, detect peak amplitudes, and, finally, approximate the peaks' greatest common divisor (Wold et al., 1996).

8. **Brightness**

Measure the amount of high-frequency content in audio signal; this is done by measuring the amount of energy above the cut-off frequency. The increment in loudness also increases the amount of high-spectrum content of a signal, thus making a sound brighter (Kim et al., 2005).

9. **Roughness**

Roughness is a way to describe the pleasantness reduction in hearing. The total roughness estimation is achieved by finding all the peaks and taking the average of all the dissonance between all the possible pairs. A typical implementation is included in the music feature extraction toolbox by Lartillot and Toiviainen (2007).

10. **Irregularity**

The spectrum irregularity measures the degree of variation in successive spectrum peaks; it can be approximated by finding the square difference in amplitude between adjacent partials. A typical implementation is included in the music feature extraction toolbox by Lartillot and Toiviainen (2007).

8.2.2 TIME-FREQUENCY DOMAIN

The concept of the time-frequency domain, sometimes called the Cepstrum domain, was introduced for the first time by Bogert et al. (1963). It is the result of taking the Fourier transform of the logarithm of the magnitude of the spectrum. Indicatively, the cepstrum can be expressed as:

$$cepestum(n) = DFT(\log(DFT(x(n)))) \tag{8.19}$$

The second Fourier transform can be replaced by the Inverse Discrete Fourier Transform (IDFT), Discrete Cosine Transform (DCT) or Inverse Discrete Cosine Transform (IDCT). The Cosine transform decorrelates the data better than the Fourier transform and, therefore, it is often preferred (Mitrović et al., 2010). The cepstrum features are good at separating convoluted components of a complex signal. Literature survey reveals that a particular type of the cepstral feature, namely Mel-frequency cepstrum coefficients (MFCCs) () are arguably the mostly popular and commonly used "universal" features for a variety of audio classification and machine audition applications.

8.2.2.1 Mel-Frequency Cepstrum Coefficients

The Mel-frequency cepstrum coefficients (MFCCs) are the most popular cepstrum-based audio features.The MFCC can represent an excellent feature vector for both speech and music signals (Kim et al, 2005), and has proven to be beneficial in the field of audio classification.

MFCC is a perceptually motivated representation; it can be defined as a short-window cepstrum of a signal. The non-linear Mel-frequency scale was developed to approximate the behaviour of the HAS. The Mel is a unit of pitch that is "the subjective impression of frequency,"

To convert a frequency f in Hertz into its equivalent in Mel, the following formula is used:

$$Pitch\,(mel) = 1127.0148 \;\; log\left(1 + \frac{f(\text{Hz})}{700}\right) \qquad (8.20)$$

Calculating MFCC

The MFCC is a short-term spectral-based feature, so the first step to calculate MFCC is to segment the audio file into overlapped small windows typically with a size of 20 to 40 ms. It is known that rectangular windowing will introduce non-trivial side lobes. To mitigate this artefact, a Hamming window may be applied on each frame.

1. Spectrum: The spectrum of each window is calculated by:
   a. Applying fast Fourier transform.
   b. Taking the absolute value to obtain the magnitude.
   c. Because of the symmetry properties of FFT, only the first half of the coefficients are taken.
2. Reducing spectrum: Now reduce the spectrum of each window to N values, where N is the size of the resulting MFCC vector; usually it will be from 12 to 20. To do that, find N Mel filter banks by following these steps:
   a. The minimum and maximum frequency need to be specified; usually the minimum frequency is zero Hertz and the maximum frequency in speech recognition systems is 4 KHz. Notice that the frequencies are in Hertz.
   b. Convert these frequencies from Hertz to Mel using Equation 8.20, and find $N + 2$ equally spaced centres in the Mel domain that start with the first centre that equals 0 and ends with $N + 2$ that equal to maximum frequency, see Figure 8.2 (a).
   c. Convert the $N + 2$ centres to Hertz using Equation 8.20; see Figure 8.2 (b).
   d. Find triangular filters for frequency centres 2 to $N + 1$, so that the triangular filter M is starting on frequency centre M, centred on frequency centre M + 1, and ends on frequency centre M. See Figure 8.2 (c).

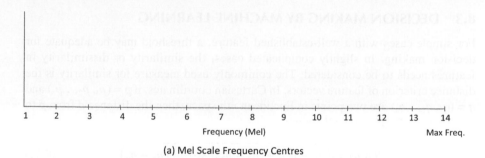

(a) Mel Scale Frequency Centres

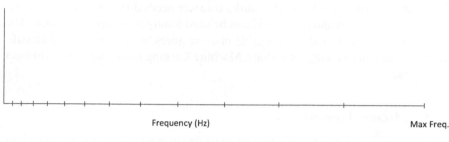

(b) Mel Scale Frequency Centres

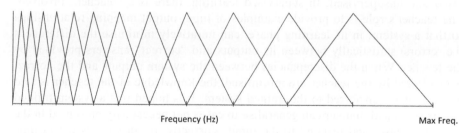

(c) 12 Mel Scale Filters.

FIGURE 8.2 Mel filters

Now Mel filter banks are ready. The next steps should be applied to the spectrum of each window. Multiply the spectrum by each filter of the N Mel filters; this will produce a vector of N elements, "one for each filter." This vector represents the reduced spectrum.

3. Log energies: Calculate the log energies of each vector using

$$Log\ Energy = log \sum_{n=1}^{W} S_n^2 \tag{8.21}$$

where W is the window size.

4. MFCC: Find the MFCC by taking the discrete cosine transform of the log energy vector.

8.3   DECISION MAKING BY MACHINE LEARNING

For simple cases with a well-established feature, a threshold may be adequate for decision making. In slightly complicated cases, the similarity or dissimilarity in features needs to be considered. The commonly used measure for similarity is the distance criterion of feature vectors. In Cartesian coordinates, if $p = (p_1, p_2, \dots p_n)$ and $q = (q_1, q_2, \dots q_n)$ are two points in Euclidean nth space, then the distance d from p to q is given by Pythagorean theorem:

$$d(p,q) = \sqrt{(q_1 - p_1)^2 + (q_2 - p_2)^2 + \cdots + (q_n - p_n)^2} \qquad (8.22)$$

For other coordinate systems, the similar measure needs different distance criteria. The distance criterion and a threshold can be used jointly to support a decision. The complexity of acoustic and audio signals often requires more sophisticated classification and decision-making algorithms. Machine learning is commonly used to meet the demands.

8.3.1   MACHINE LEARNING

Learning means adaptation of a system to its environment or expected behaviour in an iterative manner. There are two major types of learning, namely supervised and unsupervised. In supervised learning, there is a "teacher" involved. The teacher's role is to provide examples of input-output mapping relationships so that a system, in its learning phase, can iteratively minimise the differences (i.e. errors) statistically between its outputs and "correct" answers provided by the teacher. When the discrepancies between the system outputs and the correct ones offered by the teacher are minimised, the knowledge of the teacher is said to have been transferred to the trained system. It is hoped that a system trained in this statistical manner can generalise to cases not necessarily presented in the training dataset and respond to the inputs correctly. In unsupervised learning, there is no teacher involved; the learning algorithm discovers the knowledge by itself and adapts to the dataset (environment) that it is exposed to. Machine learning may be viewed as a kind of generalised or extended adaptive filter. The word "machine" here is typically a digital computer, and learning is a broad-sense adaptation. Machine learning is a vast area of study. We will use artificial neutral as an example.

8.3.2   ARTIFICIAL NEURAL NETWORK

Artificial neural networks (ANNs) are massively connectionist networks inspired by the study of the neural system in human brains. They utilise a large number of basic units to process information in a parallel manner, and they offer the capability of being trained to learn from the environment and apply the knowledge so acquired to solve problems. ANNs are studied and developed along two significantly different pathways: biological and application-driven artificial neural networks. The former is of the research interest of neuroscientists for modelling

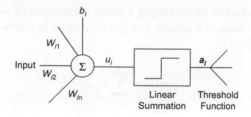

FIGURE 8.3 McCulloch and Pitts neuron model

human brains, and explaining human perceptions and behaviours. They emphasise high degrees of similarity to real brains. Application-driven neural networks, however, are deemed as inspired systems. They only need to be tied loosely to biological realities. In fact, quite different structures may be used to achieve design objectives. The ANNs used in signal processing are largely biologically inspired systems.

8.3.2.1 Neuron Models

The first neuron model was proposed by McCulloch and Pitts (1943) as a computational model of nervous activities. The McCulloch and Pitts model, as shown in Figure 8.3, is a binary device having a fixed threshold and, thus, performs simple threshold logic. A linear summation unit denoted by "sigma" calculates the sum of W_i weighted information from other neurons and a bias b_i to give the net value u_i. The gathered information u_i is then sent to a threshold function to give a bi-level output *ai*.

Over time, neuron models have evolved from the McCulloch and Pitts model to more general and sophisticated neuron models as illustrated in Figure 8.4. In this general neuron model, the linear summation unit is replaced by a basis function and the threshold is extended to a non-linear activation function. A number of different neuron models can be constructed using different basis and activation functions.

In general, the basis function $u(.)$ collects and combines input information sent by other neurons. It also intakes a bias value. Artificial neural networks can be categorised by the basis function used. Two important basis functions are linear and radial basis functions.

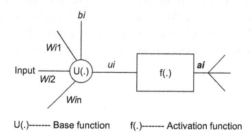

U(.)------- Base function   f(.)------- Activation function

FIGURE 8.4 General neuron model

Linear basis function (LBF) is simply a linear summation of weighted inputs and a scale value, called bias, of a neuron. For the ith neuron in a network, a linear basis function can be written as

$$u_i(\mathrm{w},\mathrm{x}) = \sum_{j=1}^{n} w_{ij}x_j + b_i \tag{8.23}$$

where the notation w_{ij} represents the connection weight from the jth neuron to the ith neuron, b_i is the bias of the ith neuron, and n is the number of connections. The LBF is used in the McCulloch and Pitts model and is still widely used in most modern neural networks.

An alternative to LBF is a hyper-sphere function known as radial basis function (RBF):

$$u_i(\mathrm{w},\mathrm{x}) = \sqrt{\sum_{j=1}^{n} \left(x_i - w_{ij}\right)^2} \tag{8.24}$$

Some insightful comparisons of these two basis functions found in classic texts (Haykin, 1999, pp. 293–240; Light 1992; Kung, 1993, p. 170) should give insights into the choice of the basis functions. In continuous linear or non-linear function approximation: there always exists an RBF-based network that accurately mimics a specified LBF-based network, or vice versa; for non-linear input-to-output mapping, LBF networks require less internal parameters than RBF networks for the same degree of accuracy; RBF networks are less numerically effective in training with back-propagation algorithm, while LBF networks are popular due to their good numerical performance; RBF networks have more constraints in architecture designs than LBF networks; RBF networks can implement arbitrarily complicated non-linear transformations, and, therefore, may solve some extremely non-linear cases not solvable by LBF networks. The above comparisons indicate that the LBF networks take several advantages over the RBF networks in solving most function mapping problems, while the RBF networks are normally reserved for extremely non-linear and extraordinarily complicated cases, which the LBF networks cannot handle.

The net value u_i of the ith neuron calculated by its basis function is sent to an activation function $f(.)$ to be further processed. The activation function is often a non-linear one, most typically a threshold or sigmoid function or an inverse tangent function. Threshold function gives a bi-level output and is often used for classification and decision making. If continuous values are expected, the sigmoid function defined by

$$a_i = \frac{1}{1+e^{-u_i}} \tag{8.25}$$

can be adopted. Alternatively, an inverse tangent function also offers continuous function regression.

8.3.3 Topology of Artificial Neural Network

The architecture of ANNs is referred to as the topological connection of neuron models to form a network. Most practical artificial neural networks for non-linear mappings and regressions are presented in a layered feed-forward fashion, but a small portion of them have added recursive structures (feedbacks are involved). Among various network architectures, multi-layer feed forward is probably the most popular architecture found in various non-linear mapping and function approximation applications.

Figure 8.5 shows a multi-layer feed forward network with two non-linear hidden layers. Multi-layered feed forward networks trained by back-propagation algorithms are generally good approximators for non-linear input to output mapping, when a sufficiently large training set is used and they also show generalisation capability (Rumelhart et al., 1986). Such networks should have at least one hidden non-linear layer to enable non-linear mapping. In theory, the more neurons and non-linear layers, the greater function mapping power a neural network possesses, but too many hidden layers may degrade training efficiency because of the presence of too many local minima in the search space (Lippmann, 1987). A two hidden non-linear layer network is adequate for approximate non-linear functions (Kung, 1993, p. 153) and so is most commonly used in various applications. The capability of such a neural network was thoroughly studied and documented by White (1990 and 1992) and Cybenco (1989).

When an LBF neuron model is embedded in a layered structure as shown in Figure 8.5 to form a neural network, the dynamic equations of a neuron model as depicted in Figure 8.4 can thus be expressed as

$$u_i = \sum_{j=1}^{N_{l-1}} w_{ij}(l)a(l-1) + b_i(l) \qquad (8.26)$$

and

$$a_i(l) = f\big(u_i(l)\big) \quad 1 \le i \le N_l; \quad 1 \le l \le L \qquad (8.27)$$

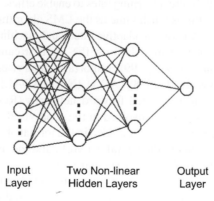

| Input Layer | Two Non-linear Hidden Layers | Output Layer |

FIGURE 8.5   An example of a multi-layer feed forward network

where l represents the layer number (the input layer is presented as layer 0), L is the number of total layers, and the input to the neural network is uniformly expressed as $a_i(0)$. Obviously, the output of the neural network is $ai(L)$.

8.3.4 SUPERVISED LEARNING RULE

An important issue in developing artificial neural networks is how they should learn. The learning algorithms are associated with architectures used, objectives of learning, and availability of external teachers.

The delta learning rule is also called Widrow-Hoff learning rule or correction learning rule (Widrow and Hoff, 1960). Consider a simple case where only one neuron, the ith neuron is present. An error signal $ei(n)$ is defined by the difference between the actual output $ai(n)$ of the neuron and the expected value $ti(n)$, where n denotes the steps of the learning process:

$$e_i(n) = a_i(n) - t_i(n) \qquad (8.28)$$

The error signal $ei(n)$ is used as a control mechanism to allow the corrective adjustment to the synaptic weights of the neuron to be made iteratively such that a cost function

$$E(n) = \frac{1}{2} e_i^2(n) \qquad (8.29)$$

is minimised. To do so, the following delta learning rule is used:

$$\Delta w_{ij}(n) = \eta e_i(n) x_j(n) \qquad (8.30)$$

and

$$w_{ij}(n+1) = w_{ij}(n) + \Delta w_{ij}(n) \qquad (8.31)$$

where the scalar η is the learning rate. The choice of the η has a significant influence on the performance of error correction learning. Too small a value results in slow learning and convergence, but learning with too large a learning rate may skip over and so miss out on important minima and solutions. In developing a neural network, much time is spent experimenting with different learning rates to enable efficient and profitable learning. The principle behind this is much same as the LMS algorithms for adaptive filters. This also highlights the link between adaptive filters and machine learning.

Supervised artificial neural models form the main stream in the development of neural networks (Kung, 1993, p. 99). Important supervised models include the perceptron model (Rowenblatt 1958), ADALINE/MADALINE model (Widrow and Hoff, 1960), and multi-layer feed forward networks (Rumelhart et al., 1986). Two phases or working modes are involved in the supervised learning: training and retrieve phases. The training data are provided in the format of example-teacher pairs, where an example is the input to the neural network, and the teacher is the expected output when the network receives the particular input.

More generally, the state of the environment is coded into input vectors being sent to the artificial neural network. The teacher is supposed to have knowledge of the environment, which is expressed in the format of input-teacher pairs or input vector-output pairs. These paired training data are the desirable input-to-output relations that need to be mapped onto the neural network. As illustrated in Figure 8.6, in the training phase, both the teacher and the neural network under training are exposed to input vectors drawn from the environment. The internal parameters of the neural network are tuned by adjusting its weights to minimise the differences between its actual outputs and the teacher values. Eventually, the ANN emulates the teacher in responding to the input vectors and, thus, the knowledge of the teacher is transferred onto the neural network.

Supervised training is an error minimisation process. The aforementioned error-correction learning rule is adopted. From a mathematical viewpoint, such learning can be viewed as a regression (curve fitting) or arbitrary function approximation problem. In practice, whenever there are teachers available, supervised training becomes the natural choice, since it, if it converges, always leads to predefined outcomes.

Supervised models can be sub-categorised into approximation-based and decision-based ones. The approximation-based networks are used to map continuous input-to-output relations. Such networks are used as the kernel processors in the acoustic parameter extractions, mapping the received speech onto continuous objective parameters. In a decision-based network, the output is bi-level, being either 1 or 0 representing "yes" or "no." Such a network is commonly used for pattern recognition or classification. The network is trained to identify whether an input belongs to a predefined class or not.

Back-propagation (Rumelhart et al., 1986; Riedmiller, 1994) is a gradient-based optimisation method and reveals attractive features of reliable convergence, tolerable training time, and good generalisation capability. The idea of training a neural network using back-propagation is to iteratively apply different input data (examples) on the input layer of the network and compare the outputs of the neural network with the teacher values (true values) to obtain the error signals. The connection weights inside the neural network are adjusted so as to minimise the

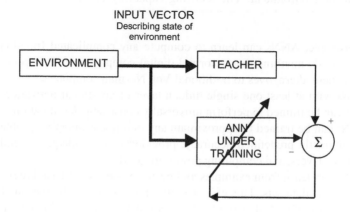

FIGURE 8.6 Supervised training (learning with a teacher)

overall errors between the outputs of the neural network and the teacher values over all the training examples. The particular rule of train the network is to update the connection weights according to the gradient-type learning formula as outlined by Equation 8.32, with the mth training sample $a^{(m)}(0)$ and corresponding teacher $t^{(m)}$ pairs so as to minimise the energy function E defined by Equation 8.33. (Note: The input layer is presented as layer 0.)

$$w_{ij}^{(m+1)}(l) = w_{ij}^{(m)}(l) + \Delta w_{ij}^{(m)}(l) \tag{8.32}$$

$$E = \frac{1}{2}\sum_{m=1}^{M}\left[t^{(m)} - out^{(m)}\right]^2 \tag{8.33}$$

The back-propagation training process follows a chain-rule

$$\Delta w_{ij}^{(m)}(l) = -\eta \frac{\partial E}{\partial w_{ij}^{(m)}(l)}$$

$$= -\eta \frac{\partial E}{\partial a_i^{(m)}(l)} \cdot \frac{\partial a_i^{(m)}(l)}{\partial w_{ij}^{(m)}(l)} \tag{8.34}$$

$$= \eta \delta_i^{(m)}(l) f'\left(u_i^{(m)}(l)\right) a_j^{(m)}(l-1)$$

where the commonly used term error signal $\delta_i^{(m)}(l)$ is defined as

$$\delta_i^{(m)}(l) = \frac{\partial E}{\partial a_i^{(m)}(l)} \tag{8.35}$$

The weight updating follows the above chain-rule from output layer towards input layer; hence the name back propagation.

The most important feature of artificial neural networks is their capability to learn from the environment. The learning capability can be further specified as follows:

1. In principle, ANNs can learn to compute any complicated functions, i.e. they can do anything that a normal digital computer can. It is possible to make them alternatives to traditional Von Norman computers.
2. ANNs with at least one single hidden layer of non-linear activation functions can be trained to perform universally consistent classification.
3. ANNs can be trained to approximate any non-linear mapping problems to an arbitrarily predefined accuracy, provided that an adequate number of neurons, layers, and computing power are used.
4. ANNs can learn from examples and perform non-model-based regressions from noisy data sets. This effective algorithm is one of the reasons for the popular use of multi-layered feed-forward networks in solving engineering problems.

Learning with artificial neural networks is a statistical method in nature in a broad sense (Haykin, 1999, pp. 84–86). The neural network methods can be viewed as statistical ones, but they differ from traditional statistics in that there is no prior assumption of a specific function, e.g. linear, exponential, certain polynomials etc. Due to the nature that certain artificial neural networks can theoretically approximate arbitrarily complicated input-output mapping relationships, the neural network model itself is data driven.

There are many other machine learning classification algorithms; examples include support vector machine, decision trees and their ensembles (e.g. Random Forests), and k-nearest neighbours algorithm. To better model stochastic nature of the signals probabilistic model such as Markov model and mixture models can be used in supervised learning. Hidden Markov model and Gaussian mixture models are the popular ones for the use in supervised learning. (Interested readers may find more information from the extended reading list.)

8.4   TRAINING, TESTING AND VALIDATION

8.4.1   TRAINING AND TESTING

It is evident that training aims to minimise the discrepancies between the system outputs and teacher's values in a minimal mean square error sense. This is fine if the learning aims to train a system to perform mapping tasks from a limited number of cases, provided that all cases are available and included in training and, afterwards, testing the system. However, machine learning as a statistical method is not generally intended for such simplistic applications. In contrast, statistical machine learning often handles data from an unrestricted space and, furthermore, "noises" are inevitably involved. For example, an ANN is trained to recognise a digitized hand-written character "a" by arbitrary writers; it is unrealistic to get all people to write an "a" and involve all people in the training set. Furthermore, each time one person writes an "a," there are some minor differences or variations. It is impossible to exhaustively obtain an infinite number of samples. Many machine learning processes literally require that in the training phase, the algorithm learns from a reasonable and limited number of cases and then generalises the knowledge to cases not necessarily included in the training. For prediction, regression, pattern recognition, and many other applications, capability of generalisation is a common concern of statistical machine learning.

Cross-validation is often used as a model validation technique to determine if a statistical analysis generalises to an independent dataset. The general idea is that some of the data are removed from the dataset before training. When training is completed, the data that were removed can be used to test the generalisation performance of the statistical model on "data not seen" in the training. There are several cross-validation methods' amongst them, the holdout cross-validation and k-fold cross-validation are the most commonly used in machine learning.

8.4.1.1 Holdout Cross-Validation

Being the simplest, the holdout method splits the dataset into two subsets, namely the training and validation sets. The learning algorithm is trained on the training set and tested (validated) with the validation set. To avoid some sequential or patterned arrangement of the original dataset, the data points should be picked randomly and assigned to training and validation sets. The holdout method does not specify the percentage of splitting (i.e. the size of the two subsets). In practice, the test set is sometimes smaller than the training set, and in a more strict sense, equal-sized training and validation subsets can be used. The drawback of the holdout method is that the evaluation shows a large variance depending on how the dataset is divided. To solve this problem, k-fold validation may be used.

8.4.1.2 K-Fold Cross-Validation

In k-fold cross-validation, the data set is first split into k subsets. The holdout method is performed k times, with one of the k subsets being used as the test set and all residuals being used for training. The evaluation is done using the average error across all k trials. The simplest k-fold cross validation is the 2-fold one. In 2-fold validation, essentially the training and validation data sets for holdout methods are exchanged and the results averaged. The most popular k-fold cross-validation is probably the 10-fold one.

8.4.2 Over-Fitting and Under-Fitting

Over-fitting means that a machine learning model represents far too closely or exactly a specific set of data; consequently, it fails to fit additional data or predict future observations reliably, i.e. compromised generalisation. An over-fitted solution in statistics often suggests that the model contains more parameters than what is necessary for the underlying structure of the data. A typical example is a statistical model that over-fits to some noise. In supervised machine learning, over-fitting often happens due to over-training. This will be discussed later.

In contrast, under-fitting means that a statistical model does not adequately represent or model the underlying structure of the data. Under-fitting often occurs when a model does not have adequate representation capability, has missed parameters, or has incorrect parameters. A typical example is the use of a straight line to fit data points showing an exponential decay process. Under-fitting can occur in supervised machine learning for a number of reasons, including inappropriate machine learning methods/models, an insufficient number of variables, under-training, and too small a training set. In addition, unsuitable step sizes and optimisation algorithms are also likely to cause under- or over-fitting problems in supervised learning.

To avoid over- and under-fitting, a supervised learning process is often continuously tested and monitored instead of using a specifically pre-set number of training iterations. Figure 8.7 illustrates typical changes of mean square errors in supervised learning when tested with a training dataset (lower curve) and a validation dataset (upper curve). As the training iteration increases, the mean square error tested with the training dataset monotonically decreases. It is interesting to note that the mean square error curve exhibits a "v" shape, when tested with the validation dataset, decreasing to reach a minimum and then increasing.

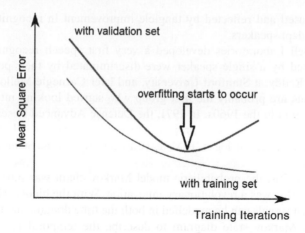

FIGURE 8.7   Illustration of over-fitting

8.4.3 STOP CRITERION, STEP SIZE, AND RESTART

Supervised learning involves iterative updating to minimise the errors similar to adaptive filters, but such a process is not "on the fly" in the retrieval phase. Some stop criteria are needed. There are a number of ways and considerations. One obvious criterion would be to stop the training when validation with the testing set reaches the minimum value. On other occasions, we can choose to stop the optimisation before a minimum is achieved, if the error is small enough for practical needs.

As mentioned in the previous section, the step size needs attention in supervised learning. The step size has an impact on a number of aspects, such as the convergence and speed of learning, and fitting issues. In general, a large step size leads to fast search of the error surface and, therefore, can potentially speed up the learning process, but when the step size is too large, the algorithm may miss the true minimum. A small step size helps with a careful search at the cost of a longer learning process. However, there is a risk that the search can get trapped in local minima.

8.5   SPEECH RECOGNITION

Automated speech recognition (ASR), which may also be called speech-to-text conversion/transcription or automated dictation, depending on the emphasis of its functionality, is a typical audio and acoustical signal pattern recognition task in a general sense and is achieved mostly through machine learning. Nevertheless, ASR has been developed into a multidisciplinary subject including computational linguistics, phonetics, semantics, data mining, and many other related subjects.

The ASR technology has developed and evolved over 60 years to a relatively mature status enabling very broad deployment on a variety of platforms and different systems for diverse applications in everyday computing experience since the first attempt to develop such systems in the 1950s at Bell Laboratories. There are several milestones symbolically represented by the pattern recognition algorithms and methods used and reflected by tangible improvement in recognition rate and capability to adapt speakers.

In 1952, Bell Laboratories developed a very first speech recognition system: 10 digits uttered by a single speaker were discriminated by their power spectra. Professor Raj Reddy at Stanford University, and later Carnegie Mellon University, and his students are probably the first group who started looking into continuous speech recognition in the 1960s. In 1971, the Defense Advanced Research Projects Agency, USA, funded a five-year project to recognize approximately 1,000 uttered words, in which BBN Technologies, IBM, and Carnegie Mellon and Stanford universities were involved.

In the late 1970s, the probabilistic model Markov chain was introduced to the ASR research. This was a revolutionary innovation. With the hidden Markov model (HMM), speech signals can be modelled in both the time domain and the frequency domains by a Markov state diagram to describe the temporal properties and a Gaussian mixture model (GMM) to characterise the spectral properties of speech signals. A very early form of the HMM for speech signal modelling was instigated by Reddy's students Jim Baker and Janet Baker at Carnegie Mellon University. The HMM-based ASR was then further developed by Fred Jelinek at IBM Research. Under Fred Jelinek's leadership, IBM developed the HMM-based ASR Tangora in the 1980s, which could handle a 20,000 word vocabulary. In 1982, Dragon Systems, one of the famous ASR software developers, was founded by James and Janet Baker. Dragon speech recognition software later became part of Nuance.

AT&T deployed the Voice Recognition Call Processing service in 1992 developed by Lawrence Rabiner's team at Bell Laboratories. Another former student of Professor Reddy, Xuedong Huang, developed the Sphinx-II system at Carnegie Mellon University and achieved outstanding performance in DARPA's 1992 assessments. Huang went on to found the speech recognition group at Microsoft in 1993. Raj Reddy's student Kai-Fu Lee joined Apple where, in 1992, he was involved in the development of Apple's speech recognition software known as Casper. There are many industry developers: Google, Microsoft, IBM, Apple, and Nuance, to name just a few.

From the 1980s to recent years, HMM/GMM-based speech models combined with feed forward artificial neural networks had been the mainstream technology for speech recognition. The current trend of continuous research and development in ASR is the use of deep learning, big data, and cloud computing. For example, a deep-learning recurrent neural network called long short-term memory (LSTM), trained by connectionist temporal classification (CTC), can significantly outperform traditional speech recognition. Recently, Google's speech recognition reportedly experienced a dramatic performance improvement through the use of CTC-LSTM.

Today's development and implementation of large-scale speech recognition systems require non-trivial resources, more than ever before. Indeed, speech recognition is a specific and multidisciplinary area. However, to ability to deploy and tailor existing technologies for specific purposes and bespoke applications has been made relatively easy thanks to the availability of many toolkits and dedicated software platforms, e.g. Microsoft HTK toolkit and Carnegie Mellon University's Sphinx software platform. Suggested texts are given at the end of this chapter for those who wish to further explore this area.

8.6   SPEAKER RECOGNITION

Automated speaker recognition should be differentiated from speech recognition. Speaker recognition aims to identify who the talker is or determine if a talker is authentically a specific person or an imposter, rather than determining what a talker is taking about.

Several established speaker recognition methods exist. These are typically classified by the feature spaces adopted and machine learning algorithms used. Mel-frequency cepstral coefficients (MFCCs), Gammatone frequency cepstral coefficients (GFCCs), and the iVectors (Dehak, 2009) are arguably the most popular features and the Gaussian mixture model (GMM) is the most commonly used machine-learning and decision-making algorithm, as suggested by a large number of publications found in an extensive literature search. Microsoft Speaker Recognition Toolbox (MSR Identity Toolbox) is written in MATLAB®. It offers a fast prototype tool to develop speaker recognition systems.

8.7   MUSIC INFORMATION RETRIEVAL

Music information retrieval (MIR) may be viewed as a branch of machine audition, i.e. the use of computers to listen to music and transcribe it back to a music score, and perform musical analysis, such as theme analysis, mood analysis, and tonal feature analysis. Music information retrieval involves audio signal analysis, machine learning and pattern recognition, musicology, and psychoacoustics, and, therefore, is truly a multidisciplinary area of study.

Typical applications of MIR include, but are not restricted to,

- Automated music transcription (Track separation and instrument recognition are often used as pre-processors.)
- Theme analysis
- Mood and expression analysis
- Automated categorisation (music genre classification)
- Automated recommendation system (music clustering)

MIR systems are based on feature extraction, feature selection, and machine learning. Psychoacoustics and musicology are heavily involved. Common features can be classified into music features (high level), and audio or signal features (low level). Commonly used signal related features include, but are not restricted to, envelope, spectrum, cepstrum, autocorrelation, centroid, spread, skewness, spectral flux, entropy, and MFCC. Music-related features include, but are not limited to, dynamic, beats, rhythm (fluctuation, beat spectrum, onset, event density, pulse clarity, etc.), timbre (attack time, attack slope, attack leap zero cross, roll off, brightness, roughness), pitch, key, and chord.

There are toolboxes available for MIR applications, e.g. the MIRtoolbox developed by Oliver Lartillot written in MATLAB and MARSYAS (music analysis, retrieval and synthesis for audio signals, an open-source framework) written in C++/Python by George Tzanetakis. With some basic signal processing knowledge and these toolboxes, one can develop a relatively simple bespoke system in a few days.

Some of the built-in audio signal feature extraction functions in MIR toolboxes can be used for audio feature extraction for general audio pattern recognition applications.

8.8 MACHINE AUDITION OF ACOUSTICS

Propagation of sound from a source to a receiver in an enclosure is often modelled as an acoustic transmission channel. Objective room acoustic parameters are commonly used to describe properties of such channels in the design and assessment of concert halls, theatres, recording studios, and many other acoustically critical spaces. Traditionally, room acoustic parameters are measured using specific testing signals such as maximum-length sequences, white noise, or sine sweeps. The use of unpleasant and noisy test signals hinders occupied and in-situ measurements. To address this issue, a number of new methods and algorithms have been developed to determine room acoustic parameters using machine audition of naturally occurring sound sources such as speech and music. In particular, reverberation time, early decay time, and speech transmission index can be estimated from received speech or music signals using statistical machine learning or maximum-likelihood estimation in a semi-blind or blind fashion. Machine audition of acoustics presented in this section is used as examples to show how machine learning and parametric models are used to solve acoustical signal processing problems.

The use of machine audition of speech or music to determine acoustic parameters of the channel through which the speech and music have been transmitted has been studied, leading to the development of a number of new algorithms and methods. Reverberation time (RT) is the most common acoustic parameter used for the assessment of room acoustics. Early decay time (EDT) was reported to better correlate with human perception of the reverberation effect and, hence, has become an ever-popular acoustic parameter. Speech transmission index (STI) is an IEC standardised objective parameter for speech intelligibility assessment in a space. These parameters are considered in this chapter. Readers who are not familiar with room acoustics may find Kuttruff (2000) a wonderful resource to learn more about these parameters. Short-time RMS values of received discrete speech utterances, in particular pronounced digits, can be used to determine RT and EDT by statistical machine learning with an ANN (Cox, Li & Darlington, 2001a; Cox & Li, 2001b). STI can be accurately determined from received running speech signals with an envelope spectrum estimator and ANNs (Li & Cox, 2001; Li & Cox, 2003). If the background noise level is reasonably low, RT and EDT can be obtained from running speech in a similar way. Speech stimuli have a limited frequency range and, therefore, can only be used to determine acoustic parameters from 250 Hz to 4 kHz octave bands effectively. Music is considered to solve the problem, but the uneven spectra of music signals mitigate the accuracy. A note-matching filter bank can be used to circumvent the problem (Kendrick et al., 2007; Kendrick et al., 2008). Envelope spectra are a relatively stable feature of anechoic speech. With the aid of eigenvector as a feature space, blind estimation of STI and RT is possible (Li & Cox, 2007).

8.8.1   Acoustic Transmission Channels and Acoustic Parameters

A received sound signal in a room is the convolution of the source and the impulse response from the source to the receiver,

$$r(t) = s(t) \otimes h(t) \tag{8.36}$$

where $r(t)$, $s(t)$, and $h(t)$ are the received signal, source signal, and impulse response of transmission channel, respectively. Room reflection models further suggest that the number of reflections increases rapidly as time elapses. In a diffuse field, temporal density of reflections in a room is given by

$$\frac{dN}{dt} = 4\pi \frac{c^3 t^2}{V} \tag{8.37}$$

where N, c, t, and V are number of reflections, sound speed, time, and volume of enclosure, respectively, rendering an extraordinarily long mixed-phase FIR filter. Hundreds of thousands of taps are often needed to accurately represent a room impulse response. Figure 8.8 (left) shows an example of a fairly short room impulse response. In idealised cases, signal energy vanishes gradually following an exponential trend.

To quantify the acoustics of a space, acoustic parameters are used. They are purposely defined objective measures that show a good correlation with human perception of sound quality or features of acoustics. All monaural objective acoustic parameters can be derived from impulse responses, while some of them can be derived from energy decay curves.

By definition, reverberation time is referred to as the period of time taken for sound pressure level to decrease by 60 dB within an enclosure after a stationary excitation is switched off abruptly. Reverberation time by this definition is referred to as RT60. As a modern trend, reverberation time is defined as the 60 dB decay time calculated by a line fit to the portion of the decay curve between −5 and −35 dB. Reverberation time, by this definition, is denoted as RT30. The use of RT30 avoids

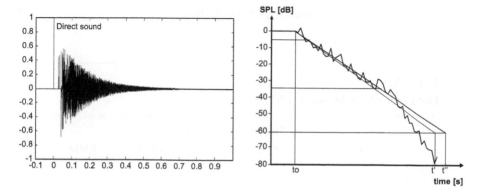

FIGURE 8.8   An example of a fairly short simulated room impulse response (left); Difference between RT60 or RT30 on a non-linear (dB scale) decay curve (right) (Figure adapted from Cox, Li, and Darlington, 2001a)

the problem of poor signal-to-noise ratio (SNR) in determining the ending point at −60 dB. However, attention should be paid to the fact that decay curves of many rooms are not exactly exponential. In the cases of non-exponential decays, RT60 and RT30 become unequal. As shown in Figure 8.8 (right), the time from t_0 to t' is RT60, while the time from t_0 to t'' is RT30. Another decay time parameter known as early decay time (EDT) is defined as the 60 dB decay time calculated by a line fitting to the 0 to −10 dB portion on the decay curve. When EDT is measured in a completely diffused field, where the decay curve in dB is linear, EDT and RT have the same value. Otherwise, EDT is found to be better correlated with subjective judgment of reverberation than RT30 or RT60. Energy decay curves can be calculated from the room impulse response using Schroeder's backwards integration method (Schroeder, 1979);

$$\tilde{h}^2(t) = \int_t^\infty h^2(x)dx = \int_0^\infty h^2(x)dx - \int_0^t h^2(x)dx \qquad (8.38)$$

The ensemble average of all possible decays $\tilde{h}^2(t)$ gives the decay curve, where $h(t)$ is the impose response. It is also worth noting that the rate of sound energy decay in a room is a function of frequency, resulting in different RTs or EDTs in octave or fractional-octave bands.

For spaces where speech communication is important, STI is often used as an objective measure for intelligibility assessment (IEC 60268-16:1998, Houtgast & Steeneken, 1973, Steeneken & Houtgast, 1980). The rationale of the STI method is that for good speech intelligibility, the envelope of speech signals should be preserved. A transmission channel degrades intelligibility by modifying (smoothening) the envelope of speech signals. The modulation transfer function (MTF) describes such envelope shaping effect and, therefore, is closely correlated with intelligibility of transmitted speech. A speech spectrum shaped noise carrier $n(t)$ is modulated by a sinusoidal function with a very low frequency (see Table 8.1):

$$m(t) = \sqrt{1 + m\cos(2\pi Ft)} \qquad (8.39)$$

TABLE 8.1

Data Point MTF Matrix for STI Extraction

(Columns: 7 octave bands from 125 to 8000 Hz; Rows: 14 modulation frequencies F from 0.63 to 12.5 Hz with 1/3 octave intervals)

	125 Hz	250 Hz	...	4 kHz	8 kHz
0.63 Hz					
0.80 Hz					
.....					
10.0 Hz					
12.5 Hz					

to generate a test signal

$$i(t) = n(t) \cdot \sqrt{1 + m\cos(2\pi Ft)} \qquad (8.40)$$

where F is the modulation frequency and m is the modulation index. The STI method uses the so-formed signals to mimic speech, and identify the envelope shaping effect and additive noise a transmission channel imposes on the signals. To do so, the test signal is applied to the input of a channel under investigation and the output is obtained. The intensity of the excitation and the response can be written as

$$I(t) = I_i[1 + m\cos(2\pi Ft)] \qquad (8.41)$$

and

$$O(t) = I_o[1 + m_o \cos 2\pi F(t - \varphi)] \qquad (8.42)$$

where m_o is the modulation index of the output intensity function and φ is time delay due to transmission. I_i and I_o are amplitudes of the corresponding sinusoidal function (mean intensities). The MFT of a channel is defined as the ratio of m_o to m as a function of modulation frequencies.

$$MTF(F) = \frac{m_o}{m} \qquad (8.43)$$

The MTF describes envelope shaping and noise injection effects and, therefore, is closely correlated with intelligibility of transmitted speech. STI is a single index calculated from 98 MTF values at 14 modulation frequencies in 7 octave bands, as illustrated in Table 8.1, following these five steps:

1. Converting $MTF(F)$ matrix into apparent S/N ratio

$$(S/N)_{app, F} = 10\log\left(\frac{MFT(F)}{1 - MFT(F)}\right) \qquad (8.44)$$

2. Limiting dynamic range to 30 dB

$$\text{If}(S/N)_{app} > 15 \text{ dB} >> (S/N)_{app} = 15 \text{ dB}$$

$$\text{If}(S/N)_{app} < -15 \text{ dB} >> (S/N)_{app} = -15 \text{ dB} \qquad (8.45)$$

$$\text{Else}(S/N)_{app} = (S/N)_{app}$$

3. Calculation of mean apparent S/N ratio

$$\overline{(S/N)}_{app} = \frac{1}{14} \sum_{F=0.63}^{12.5} (S/N)_{app, F} \qquad (8.46)$$

4. Calculation of overall mean apparent S/N by weighting the $(S/N)_{app}$, F of 7 octave bands

$$(\overline{S/N})_{app} = \sum w_k (\overline{S/N})_{app,F} \qquad (8.47)$$

The values of w_k for the 7 octave bands from 125 Hz to 8 kHz are 0.13, 0.14, 0.11, 0.12, 0.19, 0.17, and 0.14, respectively.

5. Converting to an index ranging from 0 to 1

$$STI = \frac{(\overline{S/N})_{app} + 15}{30} \qquad (8.48)$$

If the system is linear, the MTF can alternatively be calculated via the Fourier transform of the squared impulse response according to the following formula under noise free conditions (Schroeder, 1981):

$$MTF(F) = \frac{\left| \int_0^\infty h^2(t) e^{-2\pi jFt}\, dt \right|}{\int_0^\infty h^2(t)\, dt} \qquad (8.49)$$

In the presence of non-trivial ambient noise, the MTF may be determined Equation 8.49 with an additional correction factor:

$$MTF(F) = \frac{\left| \int_0^\infty h^2(t) e^{-2\pi jFt}\, dt \right|}{\int_0^\infty h^2(t)\, dt} \left(1 + 10^{(-S/N)/10}\right)^{-1} \qquad (8.50)$$

where S/N is the frequency independent signal-to-noise ratio in dB at the listener's location.

It can be seen from the definitions of these room acoustic parameters that extracting room acoustics is a system identification problem in a broad sense or a parameter estimation problem if a decay model is pre-determined. The latter is thought to be more likely to succeed than single channel blind de-convolution to obtain the complete impulse response, because acoustic parameters are dimensionally reduced down to a single value and are often related to energy decay features. In the following sections, a number of recently developed algorithms to perform such parameter estimation will be presented and discussed.

8.8.2 EXTRACTION OF REVERBERATION TIME FROM DISCRETE UTTERANCES

When a room is excited by a tone burst or noise burst, the measured short-term RMS values of the sound pressure build-up and decay caused by the switch-on and switch-off of the test signal are approximately exponential, as illustrated in Figure 8.9. The exponential decay edge gives information about the reverberation time. In fact, reverberation time extraction is simply a problem of estimating the rate of exponential decays.

If Figure 8.9 (left) was plotted in a logarithmic scale, one could simply perform a straightforward line fitting to determine the RT. Unfortunately, this is an idealised situation. The received signals of speech utterances are much more complicated and

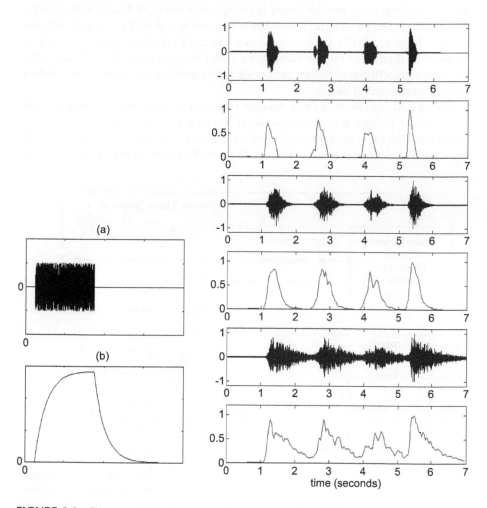

FIGURE 8.9 Exponential build-up and decay of short-time RMS value under noise burst excitation (left); Signatures of anechoic and reverberated speech utterances and their RMS envelopes (normalised amplitude vs time) (Figure adapted from Cox, Li, and Darlington, 2001a)

noisier than exponentials for a number of reasons: Speech signals are non-stationary stochastic processes, which do not have constant short-term energy like tone bursts. Individual utterances have different build-up and decay phases. Impulse responses of many rooms are not exactly exponential; impulse responses with two or more decay rates are common due to coupled spaces or non-diffused acoustic fields. A reverberation tail and the next utterance can overlap. Figure 8.9 (right) shows an anechoic and two reverberated versions of speech utterances "one, two, three, four" read by a narrator and their RMS envelopes. It can be seen from these signatures that the slopes of rise and fall edges are related to reverberation time: the longer the reverberation time, the slower the rise and decay in general. But anechoic speech utterances intrinsically have different decay slopes and they are noisy. Estimation of the reverberation time through simple examination of slopes of decay edges (in logarithmic scale) using a straightforward linear regression is simply too inaccurate to be useful. Therefore, ANNs are used to extract RTs and EDTs from received signals of pronounced digits. For the algorithm to be useful in room acoustics measurements, the design objective is to reduce estimation errors down to less than 0.1 seconds or lower than perception limens.

Figure 8.10 (top) shows a block diagram of the neural network system in its training phase. It follows a typical supervised learning regime. Reverberated speech utterances with known reverberation times are used as training examples. The training examples are pre-processed and conditioned (Figure 8.10, bottom left) to yield

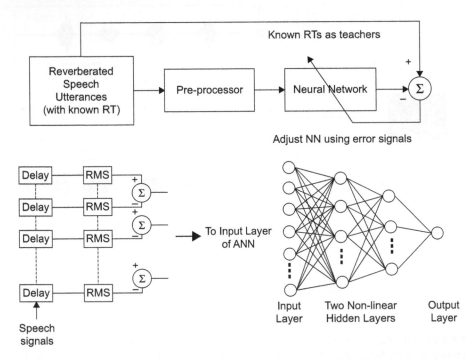

FIGURE 8.10   Neural network system, signal pre-processor, and neural network architecture (Adapted from Cox, Li, and Darlington, 2001)

suitable input vectors for the neural network. The ANN outputs and the corresponding true reverberation times (teacher) are compared to obtain the error signal. The training process is to iteratively update the internal synaptic weights of the neural network so that the mean square error between the true and the estimated reverberation times over all the training examples are minimised. In the retrieve phase, teachers and the network adjustment paths as shown in the figure are removed. Speech utterances as received by a microphone in the space are sent to the trained neural network via the same pre-processor. The trained neural network then gives accurate estimation of reverberation time.

To apply supervised neural network models to a practical problem, it is essential that the training and validation data sets be available. Real room sampling is a convincing way but it is almost impossible to obtain the required large data set across a wide range of RTs. Simulated room impulse responses are used as alternatives. Speech utterances "one," "two," and "three" are used. The convolution of the anechoic speech utterances and 10,000 simulated room impulse responses with RTs from 0.1 to 5 seconds are performed to generate a large data set. Half of it is used for training and the other half for validation. Apparently, the ANN is expected to generalise from a limited training set to arbitrary room cases that ANN has not seen in the training phase. Generalisation is an important concern, and so validation is performed strictly using the data that the ANN has not seen in the training phase. In addition, the ANN learns from examples to memorise the feature of stimulus speech signals. The data pre-processor is designed to perform four functions:

1. Normalisation of input data to total signal energy so that the presentation is independent of the input levels (not shown in the diagram).
2. Implementation of a short-term memory mechanism so that a static ANN can handle temporal speech signals.
3. Detection of the short-term average RMS value of the speech envelope.
4. Conversion of the input vector to a suitable format for the neural network by taking the difference of two adjacent RMS detectors.

Speech signals are temporal sequences; a multi-tapped delay line with short-term RMS value detectors is used to detect the short-term average energy changes of the speech signals, as illustrated in Figure 8.10 (left). For speech signals sampled at 16 kHz, 2000 point overlapped RMS detectors, providing 10 RMS values per second, are found adequate. This is equivalent to monitoring fluctuations lower than 5 Hz in a speech envelope. In this study, unfiltered speech utterances are sent to the above pre-processor. This leads to the wide-band reverberation time, which covers the frequency range of speech utterances. For octave-band reverberation time extraction, appropriate band-pass filters may be applied to speech signals and related octave-band reverberation times used as the teacher in the training phase.

A multi-layer feed-forward ANN with two hidden layers of sigmoid neurons, as depicted in Figure 8.10 (bottom, right) is adopted. Empirical comparison has found a network having 40 neurons on input layer, 20 non-linear neurons on the first hidden layer, 5 non-linear neurons on the second non-linear layer, and one linear summation output neuron is suitable. All the neurons on the non-linear layers have the same structures as the basis functions and activation functions given by Equations 8.23 and 8.25. The training follows the backwards propagation chain rule given by Equations 8.34 and 8.35. A learning rate (step size) chosen from 0.05 to 0.08 is empirically found suitable for this application, in addition to validating the trained network with the aforementioned validation set. The trained ANN system was also tested using 10 impulse responses sampled from real rooms. This not only further tested the trained network, it also validated the simulation algorithm used to generate the training and validation data sets. The errors of neural network estimation over all tests including 10 real room samples (RT from 0.5 to 2.2 seconds) are all within +/− 0.1 seconds.

Early decay time is another important objective acoustic parameter of acousticians' interest. The method described above can also be used to extract EDT from speech utterances in an almost identical way. The only difference is that the input layer of the neural network needs to be extended to have 60 neurons. This is to ensure the early decay part of the examples is sufficiently sensed. And of course, the true EDT values instead of RTs have to be used as the teacher. Testing results show that the maximum error in EDT extraction is 0.06 seconds.

Generalisation to arbitrary discrete utterances was attempted by training the ANN on more speech utterances, but results were not promising. A static neural network with a tapped delay line and RMS detectors simply would not be able to handle the complexity of arbitrary speech utterances and determine the reverberation time to the required accuracy. Thus, the application of this method is limited to the use of pre-recorded anechoic speech materials as test stimuli. Since the knowledge of speech stimuli is built in the ANN through training, during the measurement, there is no need to monitor the source. Training the ANN, though computationally burdensome, is a one-off process. In the retrieve phase, the computational load is light; RT estimates can be obtained almost immediately after the recorded speech is played.

8.8.3 Estimation of Speech Transmission Index from Running Speech

The STI, as summarised in the early part of this chapter, uses 0.63 to 12.5 Hz low frequency sinusoidal modulated noise to simulate speech signals. The modulation transfer function is used to quantify speech envelope shaping and noise interference effects, and, thus, determines speech intelligibility. If speech signals were truly sinusoidal envelope modulated white noise with equal power per frequency bin, the modulation transfer function could be easily obtained by subtraction of envelope spectra of received and original speech. For real speech signals, this relation is approximated by

$$MTF(F)(\text{dB}) \approx Ey(F)(\text{dB}) - Ex(F)(\text{dB}) \tag{8.51}$$

where $E_X(F)$ and $E_Y(F)$ are the envelope spectra of input and output long-time speech signals of a channel in decibels. The envelope spectra are obtained from squared and low-pass filtered speech signals normalised to total signal energy of the speech excerpt. The envelope spectra are typically obtained from 40–60 seconds speech excerpts to allow statistically meaningful results. Energy of envelope signals lies in a very low frequency band from immediately above DC to about 15 Hz. These

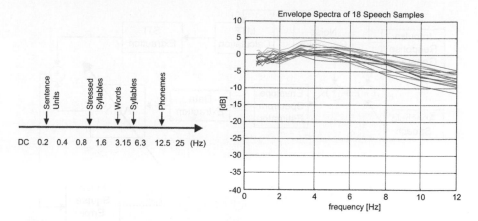

FIGURE 8.11 Components in speech envelope spectra and representative envelope spectra of running anechoic speech

frequencies are related to fluctuation of various aspects of running speech as illustrated in Figure 8.11 (left).

The envelope spectrum subtraction method shows compromised accuracy. The problem stems from the discrepancy between running speech and low frequency sine wave modulation of speech-spectrum-shaped noise. A good and suitable estimator is needed to statistically estimate envelope spectra and extract the STI. Due to the complexity of running speech, an accurate speech model is not available, making it impossible to formulate a precise analytical relationship between the envelope spectra of speech signals and MTFs in a classical sense. ANN approaches are considered to model such relations through machine learning, which is shown in Figure 8.12. It adopts the idea of envelope spectrum subtraction but uses a multi-layer feed-forward network trained by the back-propagation algorithm to memorise features of speech stimuli and generalise from a large set of impulse responses of different acoustic transmission channels. The trained neural network system can generalise to cases with impulse responses not being included in the training phase and give estimation accuracies similar to those of the standard method. (Typically, a better than 0.02 resolution in STI measurements is achievable.) Nevertheless, the model is speech excitation dependent. The information about the speech excitation is built in the network through training. Pre-recorded anechoic speech materials are needed when applying the method in field measurements.

The speech pre-processor is an envelope spectrum analyser, as depicted in Figure 8.12 (bottom). Running speech signals (60 second excerpts) are band pass filtered to obtain signals in 7 octave bands, as required by the STI method. Speech envelopes are detected using Hilbert transform. For clarity, analogue format of the equations are given.

$$ev(t) = \sqrt{s^2(t) + s_h^2(t)} \tag{8.52}$$

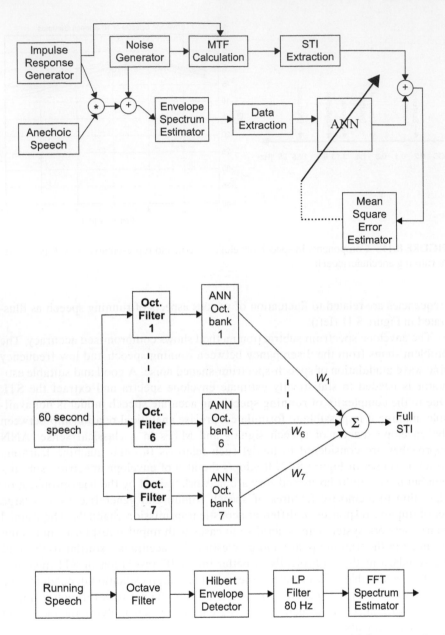

FIGURE 8.12 System architecture (top), fixed-weight network combining octave band STIs (middle), and speech envelope spectrum estimator (bottom) (Reproduced from Li & Cox, 2003)

where $s_h(t)$ is the Hilbert transform of speech signal $s(t)$ defined by

$$s_h(t) = H[s(t)] \equiv \frac{1}{\pi} \int_{-\infty}^{\infty} \frac{s(t-t')}{t'} dt' \tag{8.53}$$

Only low frequency contents found in envelope spectra are of room acoustic interest. Envelope signals are lower pass filtered by a fourth-order Butterworth filter and resample at 160 Hz. The decimated envelope signals are then passed on to an FFT power spectrum estimator to obtain envelope spectra. It is worth noting that the envelope spectra are normalised to average energy of speech signal excerpts, i.e. calibrate the spectrum analyzer so that when a sine wave having an RMS value equal to the mean intensity of the speech signal passes through, 0 dB is obtained. The normalisation has important practical and physical meanings:

- Ensuring envelope spectra are not dependent upon input signal levels,
- Expressing the frequency components of speech envelope with respect to its total (or mean) energy,
- Reflecting both speech envelope fluctuation and noise level.

Envelope spectra are re-sampled and fed into the input layer of the neural network. Not surprisingly, the window width and FFT length of the spectrum estimator has a significant impact on obtaining accurate results. According to the standard STI method, 14 data points at central frequencies of 1/3-octave bands from 0.63 to 12.5 Hz are used and found adequate if the system is trained to work with only one pre-recorded particular speech excitation. But for blind estimation (to be discussed in the next chapter), higher frequency resolutions are needed. (Hence, the envelope spectra up to 80 Hz are shown in the diagram.)

ANNs are first trained on octave band STIs, which are the values obtained following Equations 8.44–48. Among various ANNs, non-linear multi-layer feed-forward networks trained by a back-propagation algorithm, as used in the previous section, are chosen. It is empirically found that a 14-20-8-1 network performs well over all octave bands. The STI is a normalised index from 0 to 1. A hard-limit non-linear neuron at the output layer might be used to clamp the output values so that no output can possibly go beyond the interval of [0 1]. However, it is found that hard limiting the output reduces the back propagation of errors and mitigates the speed of convergence. A linear summation function without a non-linear activation function is adopted as the output neuron in the training phase, but output limiting is applied in the retrieve phase. Variable learning rates (step size) are found beneficial in this case. When an output of ANN is beyond the [0 1] interval, larger steps are used to quickly drive the ANN to produce outputs in the [0 1] region. In the final fine-tuning period, smaller steps are used. No signs of over-training were found before the maximum error in validation tests reduced to below 0.01 STI.

$$\eta = \begin{cases} (1.2 \sim 1.3)\eta & when\ output \notin [0,1] \\ \eta & others \\ (0.3 \sim 0.5)\eta & error \to 0 \end{cases} \tag{8.54}$$

Circa 14,000 simulated impulse responses covering reverberation time from 0.1 to 7 seconds are used and noise added to form the data set. Teacher values are obtained via Schroeder's method according to Equation 8.50 followed by the standard STI procedures using Equations 8.44 to 8.48. The STI by definition is considered over 7 octave bands of speech interest. The full STI is a linear combination of STIs as outline in Section 8.8.3. This can be implemented with a fixed network structure as shown in Figure 8.12 (middle).

The validation data set includes a large number of simulated examples that have not been seen in the training phase and 10 extra real room samples. Results show that the correlation coefficient of the ANN estimated and actual STIs is 0.99987 and maximum prediction error is 0.0181. Therefore, the method offers an alternative to traditional ones to accurately determine STIs in occupied spaces with naturalistic running speech.

Noting that envelope spectra are a relatively stable feature of long-time (40–60 seconds) anechoic speech signals as illustrated in Figure 8.11 (right), with the intention to achieve blind estimation of STI, training the proposed network with multiple speech excitations was attempted. To obtain the optimal estimation accuracy, envelope spectra are taken at 0.5 Hz resolution up to 80 Hz and a 160-40-20-1 network is needed. The training does converge to a relative low mean square error; nonetheless, the network does not offer sufficient accuracy (a maximum estimation error of 0.13 was found.) to replace the traditional measurement methods. More sophisticated algorithms are needed to achieve high accuracy blind estimation.

8.8.4 ESTIMATION OF REVERBERATION TIME FROM RUNNING SPEECH

Apparently, running speech is a more attractive sound source for the measurement of acoustic parameters. It is, therefore, considered to extend the above running speech STI method to the extraction of reverberation time. Recall Equation 8.50: modulation transfer function can be expressed as reverberation and noise terms. If exponential decay is assumed, this can be further written as

$$MTF(F) = \left[1 + (2\pi FRT/13.8)^2 \right]^{-1/2} \cdot \frac{1}{1 + 10^{(-S/N)/10}} \tag{8.55}$$

where F is modulation frequency and RT is reverberation time. It becomes apparent that in noise-free cases, the MTF and RT have a nonlinear one-one mapping relation. Therefore, the ANN method for STI can be used to extract RTs or EDTs in at least noise-free cases. Simulation and validation proved that this would work. If a higher than 45 dB signal-to-noise ratio is maintained, RT can be accurately determined (error < 0.1 seconds) from running speech using its envelope spectra and proposed machine-learning method (with RT or EDT as teachers in training phase). For more noisy cases, the accuracy will be compromised.

8.8.5 USING MUSIC AS STIMULI

Speech stimuli have limited frequency contents, and, therefore, can only be used to determine acoustic parameters in mid-octave bands. Experiments show in 250 Hz to 4 kHz octave bands, they have sufficient energy and can typically maintain adequate signal-to-noise ratios for room acoustic parameter extraction in most settings. Music, especially orchestral music, is considered as probe stimuli to obtain all octave band room acoustic parameters of music interest. This follows the envelope spectrum based methods outlined above, but the accuracy of estimation in mid-frequency octave bands is not as good as that from speech signals. The spectrum of a speech signal is quite "full" from 125 Hz to about 6300 Hz, with no significant discontinuity. Traditional orchestral music, however, follows equal temperament scales. Signal power is centred around discrete frequencies, each related to a note from the scale, and their harmonics. The result is a lack of excitations between notes and uneven spectra biased to particular notes in a piece (major/minor, etc.). A note-matching filter bank is developed to address this issue, as shown in Figure 8.13. For each octave band, the signal is further separated into 12 narrow frequency bands spaced according to the equal temperament scale. Envelope signals for each note are calculated and normalised to the average intensity of that note. Figure 8.13 gives an example of 12 sub-bands within 1 kHz octave band. The reverberant signal is passed though the filter bank for 12 notes where the filters' centre frequencies are determined by the equal temperament scale, starting at f#5 ($\approx$740 Hz) in the 1 kHz octave band. Envelope spectrum of the octave band is estimated from the combination of all envelope signals calculated note by note. Thus, machine leaning on envelope spectra in room acoustic parameter estimation can be extended to music stimuli. Note matching filters can mitigate the signal-to-noise-ratio problem and help improve estimation accuracy in extracting room acoustic parameters from music signals. After this treatment, over 95% of the reverberation estimates show errors of less than +/−5%, which is generally below human perception limen. It is also worth noting that standard MTF and methods derived around it may not be the best candidate for music stimuli due to the non-stationary nature of the signals. Complex modulation transfer function

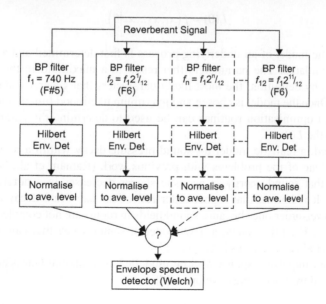

FIGURE 8.13   Envelope spectra for music signals (Reproduced from Kendrick et al., 2006)

(CMTF) has been proposed previously for music signals. Further work is needed to fully explore the potential of the CMTF in room acoustic parameter estimation.

8.9 BLIND ESTIMATION WITH A PARAMETRIC MODEL: MAXIMUM LIKELIHOOD ESTIMATION

Blind estimation of reverberation time or early decay time can be achieved using maximum likelihood estimation (MLE) of decay phases found in running speech or music signals based on a suitably chosen decay model. Strictly speaking, this is not a machine learning method but a parametric model based regression approach. Given the relevance of machine audition of acoustics, this method is included in this chapter. Reverberation time was estimated by performing an MLE on decays following speech utterances with an exponential decay model (Ratnam et al., 2003). However, the idealised exponential model is based on the assumption of a completely diffused field. Non-exponential or weakly non-exponential decays are commonplace in many rooms. This is why other reverberation parameters, most notably the early decay time, are defined. The single exponential model limits the accuracy of estimation. More sophisticated decay models, signal segmentation, and selection can improve the accuracy in blind room acoustic parameter extraction, reducing errors down to below perceptual difference limen (Kendrick et al., 2007).

MLE is a method used for parametric estimation in statistics. In essence, if there exists a parametric model for a statistical process, in the form of a probability density function f, then the probability that a particular set of parameters θ are the parameters that generated a set of observed data $x_1, x_2, \ldots x_n$, is known as the likelihood L. This is denoted by

$$L(\theta) = f\left(x_1, x_2 \ldots x_n \,|\, \theta\right) \qquad (8.56)$$

An analytic model of an underlying process is first determined and a likelihood function formulated. The parameters that result in a maximum in the likelihood function are, by definition, the most likely parameters that generated the observed set of data. Once the model is chosen and maximum likelihood function formulated, many canned optimisation routines can be used to determine the parameter(s) by maximising the $L(\theta)$.

To succeed, a realistic model for sound pressure decay within a room needs to be defined. One of the problems with previous work (Ratnam et al., 2003) is the assumption that sound energy in acoustic spaces decreases in a purely exponential fashion. It is known that higher-order reflections can often decay at different rates than lower-order ones because sound fields in rooms are not completely diffuse in most cases. For this reason, a new model of sound decay that can account for non-exponential decay curves is proposed.

Let a room impulse response $h[n]$ be modelled as a random Gaussian sequence $r[n]$ modulated by a decaying envelope, $e[n]$.

$$h[n] = e[n]r[n] \qquad (8.57)$$

where n is the sample number. The envelope is represented by a sum of exponentials:

$$e[n] = \sum_{k=1}^{M} \alpha_k a_k^n \qquad (8.58)$$

where a_k represent decay rates, α_k are $e[n] = \alpha a_1^n + (1-\alpha)a_2^n$ weighting factors, and M is the number of decays. If two decay rates are chosen, it can be weighted by a single factor:

$$e[n] = \alpha a_1^n + (1-\alpha)a_2^n \qquad (8.59)$$

where a_1 and a_2 represent the two decay rates and α is a weighting factor that changes the level of contribution from each individual decay. This enables the representation of an energy response with a non-uniform decay rate and, by changing α, the model can adapt to best fit the decay phases. Figure 8.14 shows the sum of two exponentials models with different decay rates in early and late parts on a decay curve, where the factor α acts to define a knee point where the influence of the two decay rates cross over from a_1 to a_2. Decay curves obtained from three different weighting factors are illustrated.

More exponentials, as formulated in Equation 8.58, can be used to model the decay but at the cost of extra computational overhead when optimising the likelihood function. However, as the end result of this processing is to calculate reverberation parameters that are necessarily "averaged" over a considerable amount of the decay, using a great number of exponentials and trying to fit the fine detail of the decay is unnecessary. It is empirically found in the current study that the setting with two decay rates is sufficient in the vast majority of room cases. The likelihood of a sequence of independent, identically distributed, Gaussian variables occurring is given by (Weisstein, 2006)

$$L(r,\sigma,\mu) = \prod_{n=0}^{N-1} \frac{1}{\sqrt{2\pi}\sigma} e^{\left(-\frac{(r[n]-\mu)^2}{2\sigma^2}\right)} \qquad (8.60)$$

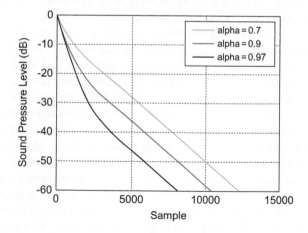

FIGURE 8.14 A dual-exponential decay model with decay curves obtained from three different weighting factors (Reproduced from Kendrick et al. 2007)

where μ is the mean and σ^2 the variance of the Gaussian process. The room impulse response model has no DC component, so $\mu = 0$. For the decay phases found in reverberated sounds s, the envelope is of interest. Thus, the probability of the sequence, which has a zero mean and is modulated by an envelope e, is given by

$$L(s;\sigma,e) = \prod_{n=0}^{N-1} \frac{1}{\sqrt{2\pi e[n]\sigma}} e^{\left(-\frac{s[n]^2}{2e[n]^2\sigma^2}\right)}$$ (8.61)

This can be rearranged to give:

$$L(s,\sigma,e) = e^{\left(\sum_{n=0}^{N-1}\frac{-s[n]^2}{2e[n]^2\sigma^2}\right)} \left(\frac{1}{2\pi\sigma^2}\right)^{N/2} \prod_{n=0}^{N-1} \frac{1}{e[n]}$$ (8.62)

The proposed decay model Equation 8.59 is substituted into Equation 8.62. It is more convenient to work with a logarithmic likelihood function, since the multiplication becomes summation. The log likelihood function is

$$\ln\{L(s,\sigma,a_1,a_2,\alpha)\} = -\sum_{n=0}^{N-1} \frac{\left[\alpha a_1^n + (1-\alpha)a_2^n\right]^{-2} s[n]^2}{2\sigma^2}$$

$$-N/2 \ln\left(2\pi\sigma^2\right) - \sum_{n=0}^{N-1} \ln\left[\alpha a_1^n + (1-\alpha)a_2^n\right]$$ (8.63)

As a Gaussian process has been assumed, maximising the log likelihood function with respect to the decay parameters α, a_1, and a_2 yields the most likely values for these parameters. This is achieved by minimising the minus log-likelihood function. Many existing algorithms or canned routines can perform the required optimisation. The sequential quadratic programming (SQP) type of algorithm is found suitable for this application (Fletcher, 2000). Once the parameters in Equation 8.63 are determined, the decay curve is obtained. Room acoustic parameters can be calculated according to their standard definitions from the decay curve. In addition to blind estimation, another attractive feature of this new and improved MLE method is its ability to obtain decay curves, which are useful in diagnosing acoustic problems.

The dual-decay model MLE method for RT and EDT extractions has been tested rigorously using anechoic speech and music excerpts convolved with 20 real room sampled impulse responses and 100 simulated impulse responses. The method relies on exploiting the free decay phases found in speech or music. Speech signals provide an ideal excitation in the sense that they contain many free decays for analysis. Typically, a 90-second excerpt of running speech from an untrained narrator is sufficient for accurate estimations. When speech stimuli are used, the dual decay MLE algorithm shows reduced prediction errors to below subjective difference limens in most cases, hence providing a useful tool for measuring rooms. However, the limited bandwidth of speech stimulus causes problems at low frequencies. When the algorithm is applied to receive orchestral music signals, however, the estimates tend to be less accurate than those obtained from speech. This is not surprising, because music often lacks free decay phases sufficiently long for reverberation time estimation. The accuracy to some extent depends upon the style of music used. Details about estimation errors of speech and music based measurements using the dual-decay MLE method can be found in Kendrick et al., 2007.

Estimation of acoustic parameters of sound transmission channels via machine audition of transmitted speech or music signals has many potential applications. It enables the acquisition of occupied acoustic data of spaces such as concert halls and theatres. It also facilitates in-situ acoustic channel modelling and, subsequently, provides information for online/real-time channel equalisation. When blind estimation is applied, a signal pre-processor might be designed to compensate and equalise channel distortions, so improving the performance of further machine auditions such as automated speech recognitions and automated music score transcription. Last but not least, acoustic parameters have been developed and tested extensively over 100 years. It is known that they have good correlation to perceived sound characteristics or acoustic features. To some extent, machine audition to obtain acoustic parameters might be viewed from a perspective of cognitive science cognitive as mimicking certain low levels of perception using a computer.

SUMMARY

Machine learning as a branch of artificial intelligence in computer science has permeated modern audio and acoustic signal processing. The fusion of machine learning and signal processing has led to the new era of audio signal processing or, more precisely, audio informatics, in which the outcomes of audio signal processing may take a format other than audio—speech-to-text conversion, for example. Some machine learning algorithms may be viewed as a variation or broad sense adaptive filters. Following a summary of some commonly used features of audio signals, this chapter presented the concept of supervised learning with the Widrow-Hoff rule. This chapter then discussed multi-layered and feed-forward artificial neural networks trained with the back-propagation algorithm. As worked examples, this chapter also discussed how room acoustic parameters were extracted from received audio signals.

REFERENCES

Bogert, B. P., Heacy, M., and Tukey, J. W. (1963) "The Quefrency Alanysis of Time Series for Echoes: Cepstrum, Pseudo Autocovariance, Cross-Cepstrum and Saphe Cracking," *Proceedings of the Symposium on Time Series Analysis,* pp. 209–243.

Burred, J. J. and Lerch, A. (2003) "A Hierarchical Approach to Automatic Musical Genre Classification," *In Proceedings of the 6th International Conference on Digital Audio Effects (DAFx-03),* London, UK, September 8–11, 2003.

Cox, T. J., Li, F. F., and Darlington, P. (2001a) "Extraction of Room Reverberation Time from Speech Using Artificial Neural Networks," *Journal of the Audio Engineering Society,* Vol. 49, No. 4, pp. 219–230.

Cox, T. J. and Li, F. F. (2001b) "Using Artificial Intelligence to Enable Occupied Measurements of Concert Hall Parameters," ProcEedings of the 17th ICA, 3A.08.05., Italy.

Cybenco, G. (1989) "Approximation by Superpositions of a Sigmoidal Function, Mathematics of Control," *Signals and Systems*, Vol. 2, pp. 303–314.

Dehak N., Dehak, R., Kenny P, Brummer, N., Ouellet P., and Dumouchel P. (2009) "Support Vector Machines versus Fast Scoring in the Low-Dimensional Total Variability Space for Speaker Verification," Proceedings of *INTERSPEECH 2009*, pp. 1559–1562, Brighton, UK.

Fletcher, R. (2007) "The Sequential Quadratic Programming Method, Nonlinear Optimization," pp. 165–214, *Part of the Lecture Notes in Mathematics book series*, Springer.

Gerhard, B. D. (2003) "Computationally Measurable Temporal Difference between Speech and Song," *PhD thesis*, Simon Fraser University.

Houtgast T. and Steeneken, H. J. M. (1973) "The Modulation Transfer Function in Room Acoustics as a Predictor of Speech Intelligibility, *Acustica*, Vol. 28, 1973, pp. 66–73.

IEC 60268-16:1998 (1998) "Sound System Equipment, Part 16: Objective Rating of Speech Intelligibility by Speech Transmission Index."

Kendrick, P., Li, F. F., Cox, T. J., Zhang, Y., and Chambers, J. A. (2007) "Blind Estimation of Reverberation Parameters for Non-Diffuse Rooms," *Acta Acustica*, Vol. 93, No. 5, pp. 760–770.

Kendrick, P., Cox, T. J., Li, F. F., Zhang, Y., and Chambers, J. A. (2008) "Monaural Room Acoustic Parameters from Music and Speech," *Journal of the Acoustical Society of America*, Vol. 124, Issue 1, pp. 278–287.

Lartillot, O. and Toiviainen, P. (2007) "A Matlab Toolbox for Musical Feature Extraction From Audio," *Proceedings of the 10th International Conference on Digital Audio Effects (DAFx-07)*, Bordeaux, France.

Li, F. F. and Cox T. J. (2001) "Extraction of Speech Transmission Index from Speech Signals Using Artificial Neural Networks," *110th AES Convention*, Amsterdam, paper 5354.

Li F. F. and Cox T. J. (2003) "Speech Transmission Index from Running Speech: A Neural Network Approach," *Journal of the Acoustical Society of America*, Vol. 113, Issue 4, pp. 1999–2008.

Li F. F. and Cox T. J. (2007) "A Neural Network Model for Speech Intelligibility Quantification," *Journal of Applied Soft Computing*, Vol. 7, Issue 1, pp. 145–155.

Lippmann, R. P. (1987) "An Introduction to Computing with Neural Nets," *IEEE ASSP Magazine*, 4: 4-22, pp. 153–159.

Lu, L., Zhang, H., and Jiang, H. (2002) "Content Analysis for Audio Classification and Segmentation," *IEEE Tranactionss on Speech and Audio Processing*, Vol. 10, No. 7, pp. 504–516.

McCulloch W. S. and Pitts W. (1943) "A Logical Calculus of the Idea Immanent in Nervous Activities," *Bulletin of Mathematical Biophysics*, Vol. 5, pp. 115–133.

Mitrović, D., Zeppelzauer, M., and Breiteneder, C. (2010) "Chapter 3—Features for Content-Based Audio Retrieval" in V. Z. Marvin (Ed.), *Advances in Computers* (Vol. 78, pp. 71–150): Elsevier.

Ratnam, R., Jones, D. L., Wheeler, B. C., O'Brien, W. D. Jr, Lansing, C. R., and Feng, A. S. (2003) "Blind Estimation of Reverberation Time," *Journal of the Acoustical Society of America*, Vol. 114, pp. 2877–2892.

Riedmiller, M. (1994) "Supervised Learning in Multi-Layer Perceptrons—From Back Propagation to Adaptive Algorithms," *International Journal of Computer Standards and Interfaces*, Vol. 16, pp. 27–35.

Rowenblatt F. (1958) "The perceptron: A Probabilistic Model for Information Storage and Organisation in the Brain," *Psychological Review*, Vol. 65, pp. 386–408.

Rumelhart, D. E., Hinton, G. E., and Williams, R. J. (1986) "Learning Internal Representations by Error Propagations," in *Parallel Distributed Processing: Exploration in the Microstructure of Cognition*, MIT Press, Cambridge, MA., Vol. 1, pp. 318–362.

Scheirer, E., and Slaney, M. (1997) "Construction and Evaluation of a Robust Multifeature Speech/Music Discriminator," *In Proceedings of the IEEE ICASSP-97*, Vol. 2, pp. 1331–1334.

Schroeder, M. R. (1979) "Integrated Impulse Method Measuring Sound Decay without Impulses, *Journal of the Acoustical Society of America*, Vol. 66, pp. 497–500.

Schroeder, M. (1981) "Modulation Transfer Functions: Definition and Measurement," *Acustica*, Vol. 49, pp. 179–182.

Steeneken H. J. M. and Houtgast T. (1980) "A Physical Method for Measuring Speech Transmission Quality," *Journal of the Acoustical Society of America* Vol. 67, No. 1, pp. 318–326.

Weisstein, E. (2006) *Wolfram Mathworld: Eric Weisstein.*

White, H. (1990) "Connectionist Nonparametric Regression: Multilayer Feedforward Networks Can Learn Arbitrary Mappings," *Neural Networks*, Vol. 3, pp. 535–550.

White, H., and Gallant, A. R. (1992) "On Learning the Derivatives of an Unknown Mapping with Multilayer Feedforward Networks, *Neural Networks*, Vol. 5, pp. 129–138.

Widrow, G. and Hoff, M. E. (1960) "Adaptive Switching Circuit," *IRE Western Electronic Show and Convention: Convention record*, pp. 96–104.

Wold, E., Blum, T., Keislar, D., and Wheaton, J. (1996) "Content-Based Classification, Search, and Retrieval of Audio." *IEEE MultiMedia*, Vol. 3(3), pp. 27–36.

BIBLIOGRAPHY AND EXTENDED READING

For audio features:

Kim, H-G., Moreau, N., and Sikora, T. (2005) *MPEG-7 Audio and Beyond: Audio Content Indexing and Retrieval*, Wiley.

For artificial neural networks:

Heaton, J. (2015) *Artificial Intelligence for Humans, Volume 3: Deep Learning and Neural Networks*, CreateSpace Independent Publishing Platform.

Light, W. (1992) "Ridge Functions, Sigmoidal Functions and Neural Networks," in Cheney, E. W., Chui, C. K., and Schumaker, L. L. eds., *Approximation Theory* (VII), pp. 163–190, Kluwer Academic Publishers, Boston, USA.

Kung, S. Y. (1993) *Digital Neural Network, Information and System Science Series.* Prentice-Hall Inc., Englewood Cliffs, New Jersey, USA.

Haykin, S. (1999) *Neural Networks: A Comprehensive Foundation*, Second Edition, Prentice Hall, Upper Saddle River, New Jersey, USA.

Haykin, S. (2008) *Neural Networks and Learning Machines: A Comprehensive Foundation*, 3rd edition, Pearson.

Heaton, J. (2015) "Artificial Intelligence for Humans, Volume 3" *Deep Learning and Neural Networks*, CreateSpace Independent Publishing Platform.

For room acoustics:

Kuttruff, H. (2000) *Room Acoustics*, Fourth Edition, Spon Press, London.

RECOMMENDED SOFTWARE AND TOOL KITS

Typically, music and speech processing software platforms are built upon a collection of feature extraction functions and a few machine learning engines. They provide fast prototyping utilities for many audio signal processing systems.

Music Information Retrieval

MIRtoolbox by Olivier Lartillot et al.

https://www.jyu.fi/hytk/fi/laitokset/mutku/en/research/materials/mirtoolbox

This is a collection of MATLAB functions and is available free of charge under the GNU General Public License. MIRtoolbox requires MATLAB Signal Processing toolbox.

Marsyas (Music Analysis, Retrieval and Synthesis for Audio Signals) by George Tzanetakis

http://marsyas.info/

A powerful audio signal processing toolkit with a specific emphasis on music information retrieval, distributed under the GNU Public Licence (GPL) Version 2. Marsyas has been used in a range of projects by leading industries.

Speech Recognition

CMUSphinx

https://cmusphinx.github.io/

A well-established and application-oriented speech recognition toolkit that can even run on mobile devices; BSD-like licence.

HTK Toolkit (Hidden Markov Model Toolkit)

http://htk.eng.cam.ac.uk/

A set of software for implementing HMMs. Developed for speech recognition, but can and has been used for many other applications using the HMMs.

Speaker Recognition

Microsoft MSR Identity Toolbox

https://www.microsoft.com/en-us/download/details.aspx?id=52279

A MATLAB toolbox for speaker recognition research.

EXPLORATION AND MINI PROJECTS

Exploration

1. Download the file *ANN.zip*, unpack it in working directory of MATLAB. This zipped file contains a few heavily commented example codes for non-linear relation mapping, and is built on a multilayer feed-forward annual network trained by the back-propagation algorithm. Read these codes and try to execute them. The codes can later be used as "blueprints" or a starting point for the suggested mini project.

2. Modify learning rate, number of neurons and number of layers, number of iterations, etc., and observer the behaviour of the neural network.

3. Set some stop criteria to end the training with justifiable reasons.

4. Add some noises to the inputs; divide the dataset into training and validation subsets; adjust parameters and the size (step size, number of neurons, and layers) of the neural network so that over-fitting can be observed. Expand the code to implement a stopping criterion so that the training stops immediately before over-fitting occurs.

MINI PROJECT IDEAS

- Collect some pure audio clips of music and speech to form data sets. Train an ANN to classify the audio clips into music or speech from 5-second excerpts of these clips, using 2 to 3 suitable features. Validate the trained classifier.

- As a step further, consider data sets containing pure music, pure speech, and overlapped music and speech. Design and implement a system that can categorise the audio clips into these three classes. Note that for complicated classification problems such as this particular task, more audio features may be needed.

- Audio classification can be extended to many other applications, e.g. classification of male and female voice, and detection of faulty machinery from abnormal noise during operation.

More similar mini project ideas and application scenarios can be proposed.

9 Unsupervised Learning and Blind Source Separation

Machine learning does not always need a teacher. Unsupervised learning is another form of learning mechanism. Unsupervised learning is often used in signal processing information extraction, dataset dimensionality reduction, signal separation, and many more situations. Unsupervised learning has no teachers involved, so it does not necessarily yield the expected outputs. Instead, an unsupervised network, once trained, forms internal representations for encoding features of the input and, therefore, creates new classes automatically. (Becker, 1991) In layperson's words, unsupervised learning can discover knowledge. Furthermore, unsupervised networks can be viewed as information encoders and can perform data reduction to extract most significant information. Although unsupervised learning may lead to outcomes without significant physical meanings, the behaviour of some well-structured unsupervised models are known. They have links to certain statistical methods, providing an important means to perform data mining and compression. For example, unsupervised learning can be used to perform principal components analysis (PCA) and independent component analysis (ICA).

Digital audio and files are big data. The raw data carry much information that may not be important or of our interest in some applications. We intend to reduce the dimensionality and find some most significant or dominant components for specific purposes. For example, if we want to understand the dynamic range of a recording, we do not need to know information about pitch, chord timbre, and many other irrelevant aspects. Dimensionality reduction itself may function as a kind of feature extraction or help with feature extraction, and reduce the data points to a more manageable size for further processing in machine audition.

On other occasions, two wanted signals are mixed up. We are interested in both of them. This is a separation problem instead of a filtering one. Separation problems may be converted into two or more filtering problems if the two signals span over different frequency ranges. Cross-over filters in loudspeakers or active signal level cross-over boxes are typical examples. However, when signals to be separated span the same or similar frequency range, traditional filters operating in the Fourier spectral domain will fail. A typical example is the famous cocktail party effect problem, which intends to separate two arbitrarily mixed speech signals. If we have neither information about how the signals are mixed nor prior knowledge of the signals, this is called a blind source separation problem.

9.1   HEBBIAN LEARNING (SELF-ORGANISED LEARNING)

Hebb postulated a learning mechanism known as Hebbian learning. According to Hebb, "When an axon of cell A is near enough to excite a cell B and repeatedly or persistently takes part in firing it, some growth process or metabolic changes take place in one or both cell such that A's efficiency as one of the cells firing B, is increased." (Hebb, 1949) Hebb's postulation is the oldest but probably most famous one in the sense that interaction of neurons in a Hebbian manner would automatically lead to unsupervised learning. Hebb's postulation was further developed and proved by a number of researchers. Changeux and Danchin (1976) rephrased and expanded Hebb's original statement. Sejnowski (1977,a,b) confirmed the convergence of self-organised learning. Brown et al. (1990) identified four important properties of Hebbian learning. Eggermont (1990) identified and explained that Hebbian learning is, in fact, a correlation mechanism.

For a general neuron in a neural network, as outlined in Chapter 8, Hebbian learning can be formulated as

$$\Delta w_{ij}(n) = \eta a_i(n) x_j(n) \tag{9.1}$$

Consequently, when the input is highly correlated with the output, it has significant importance and should be enhanced. As the learning, i.e. adaptation of weights, is determined by the input and output, no teacher value is needed; hence, Hebbian learning is associated with unsupervised neural networks.

Unsupervised models are deemed as being self-organising; they learn from underplaying statistics from the dataset. Two important unsupervised neural network models, namely principal component analysis (PCA) and independent component analysis (ICA) neural networks, are considered in the subsequent sections. The PCA networks have much in common with normal statistical PCA and are found commonly in today's deep learning algorithms as input stages or earlier layer(s). ICA-based blind signal separation and de-convolution (BSSD) techniques show potential usefulness in modelling the cocktail party effect and separating signals that are in the same frequency range but statistically independent.

9.2   PCA NEURAL NETWORKS

In unsupervised learning, neural networks learn from the environment by looking at the data presented to them. For most varieties of unsupervised learning, the targets are the same as the inputs. Therefore, unsupervised learning is usually used to compress input information (Deco and Obradovic, 1996): An unsupervised neural network identifies the properties of its input data and learns to reflect these properties in either its output or internal parameters. The properties of input data that the neural network focuses on and the compressed formats that the neural network takes depend on the learning strategy and the network architecture used. The eigen-filter type of PCA is a typical example.

The concept of PCA is not new in statistics and is used to identify the most representative low dimensional subspaces from a high-dimensional vector space. In other words, PCA can reduce variable numbers while maintaining the most important feature of observed data. But performing PCA on a large data set is known to be complicated. Currently used standard computational algorithms are the so-called canned routines (Parlett, 1998) and are available from several special statistical and numerical packages. They involve finding the eigenvectors and eigenvalues of the correlation matrix of a data set. Extracting principal components is, in fact, a variance maximising rotation of the original variable space. An eigen-component of a stochastic process is its projection on the one-dimensional subspace spanned by the corresponding eigenvector. The first m principal components are the first largest m eigen-components that include the most variance. A common problem in statistical pattern recognition is feature selection. Feature selection refers to the process that data space is transformed into a feature space. The transformation is designed in such a way that the most representative or intrinsic information is retained while the less significant information is discarded. As a result, PCA finds its application in such a scenario. A simple type of unsupervised neural network is closely associated with PCA and, therefore, is sometimes called a PCA neural network.

9.2.1 HEBBIAN MAXIMUM EIGENFILTER AND PCA

When the Hebbian learning rule is performed on a single linear neuron, as shown in Figure 9.1, the synaptic weights w_i will converge to a filter for the first principal component of the input distribution. The neural network used here is very simple. It has only one linear neuron, which means a summation node without an activation function.

The dynamic equation of the neural network is

$$y(n) = \sum_{i=1}^{m} w_i x_i \qquad (9.2)$$

where n represents nth sample of the observation data.

The Hebbian learning rule discussed in the previous section is applied, so

$$w_i(n+1) = w_i(n) + \eta y(n) x_i(n) \qquad (9.3)$$

where η is the learning rate. Such a learning rule may cause unlimited growth of synaptic weights. Certain saturation or normalisation is often needed in training the

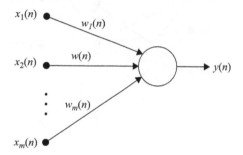

FIGURE 9.1 Maximum eigenfilter PCA neural network

neural network. One possible and also commonly used way to perform the normalisation is described by Equation 9.4:

$$w_i(n+1) = \frac{w_i(n) + \eta y(n)x_i(n)}{\sqrt{\sum_{i=1}^{m}\left[w_i(n) + \eta y(n)x_i(n)\right]^2}} \tag{9.4}$$

Once the network is converged, the synaptic weights reach steady-state values. They are the eigenvectors for the first principal component, and the variance of the outputs of the network is the maximum eigenvalue. Therefore, this kind of simple "network" with only one neuron trained on the dataset itself with a normalised Hebbian learning updating formula is called the maximum eigenfilter.

9.2.2 GENERALISED HEBBIAN ALGORITHM AND PCA NETWORK

The Hebbian maximum eigenfilter gives the first principal component. When more principal components are needed, the network structure is extended as shown in Figure 9.2 and the generalised Hebbian algorithm (GHA) can be used to obtain all needed principal components.

The PCA neural network has only one linear processing layer. The network has m inputs and l outputs, and the number of outputs is always fewer than that of inputs. The trainable parameters are synaptic weights w_{ij}, which connect the ith input to the jth neuron, where $i = 1, 2, 3, \ldots m$ and $j = 1, 2, 3, \ldots l$. The dynamic equation of such a network can be written as

$$y_j(n) = \sum_{i=1}^{m} w_{ji}(n)x_i(n) \tag{9.5}$$

and the GHA updating equation is

$$\Delta w_{ij}(n) = \eta\left[y_j(n)x_i(n) - y_j(n)\sum_{k=1}^{j} w_{ki}(n)y_k(n)\right] \tag{9.6}$$

Once training reaches convergence, weight w_{ij} of neuron j converges to the ith component of the eigenvector related to the jth eigenvalue of the correlation matrix of

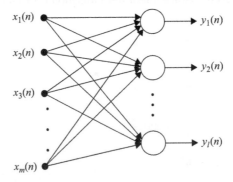

FIGURE 9.2   Hebbian learning PCA neural network

the inputs. The variances of outputs are the first lth maximum eigenvalues. Obviously, the Hebbian maximum eigenfilter is a special case of the GHA, where $l = 1$.

9.3 ICA NEURAL NETWORKS AND BLIND SOURCE SEPARATION

ICA neural networks, evolved from PCA networks, are unsupervised models for separating statistically independent signals and were previously applied to speech separation problems by Bell and Sejnowski (1995). The ICA neural network offers one possible solution to the so-called blind source separation (BSS) problems. ICA networks can also be used to perform blind de-convolution in the frequency domain, which means separating convoluted signals without prior knowledge of the original signals. Objective acoustic parameter extraction is essentially a de-convolution problem and the source-independent measurement is associated with the blind de-convolution. As the formulation of ICA network starts from the assumption of two independent sources via two different transmission channels, it is only suitable for multichannel (at least two channels) BSS.

The learning objective of ICA neural networks is to minimise the mutual information between the outputs as illustrated in Figure 9.3. The most typical scenario of the ICA neural network is to separate independent sources, say speech and noises, which is effectively BSS. In a two-channel case, the BSS can be described as shown in Figure 9.3.

Figure 9.3 illustrates the mixing and blind de-mixing process in a matrix format. Two independent sources, s_1 and s_2, are linearly mixed by arbitrary coefficients a_{11}, a_{12}, a_{21}, and a_{22} to give the mixture x_1 and x_2 according to the left half in Figure 9.3. This can be written in a matrix format

$$\mathbf{X} = \mathbf{AS} \tag{9.7}$$

where the input vector $\mathbf{S} = [s_1, s_2]^T$, the mixture vector $\mathbf{X} = [x_1, x_2]^T$, and the mixing matrix $\mathbf{A}$ is

$$\mathbf{A} = \begin{bmatrix} a_{11} & a_{12} \\ a_{21} & a_{22} \end{bmatrix} \tag{9.8}$$

and $\mathbf{A}$ is non-singular.

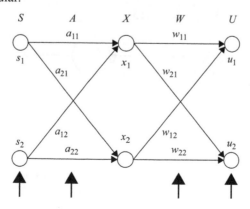

FIGURE 9.3 Defining the mixing and de-mixing in a matrix format

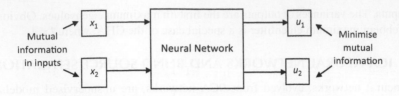

FIGURE 9.4 Training objective of ICA neural networks

The BBS is to find a de-mixing matrix $\mathbf{W}$ as shown in the right half of Figure 9.3 so that u_1 and u_2, which are recovered versions of s_1 and s_2, can be obtained by

$$U = WX \tag{9.9}$$

where the de-mixing matrix is

$$\mathbf{W} = \begin{bmatrix} w_{11} & w_{12} \\ w_{21} & w_{22} \end{bmatrix}, \tag{9.10}$$

and the recovered vector is

$$U = \begin{bmatrix} u_1, u_2 \end{bmatrix}^T. \tag{9.11}$$

This is not a simple matter of finding the inverse matrix $\mathbf{W}$ for a known matrix $\mathbf{A}$. BSS by definition means that the mixing coefficients are arbitrary, i.e. unknown though fixed. Statistical methods are sought to determine the de-mixing matrix $\mathbf{W}$.

Since sources s_1 and s_2 are assumed to be independent, if they are separated successfully, then the outcomes u_1 and u_2 should be independent. This can be achieved by minimising mutual information found in u_1 and u_2 using an unsupervised neural network, as depicted in Figure 9.4, with only one linear summation layer, as depicted in Figure 9.5.

The two neurons have linear summation basis functions as used in all the neural networks in the previous chapters, but may have different types of activation functions. Madhuranath and Haykin (1998) proposed the following activation function, which is reported superior to others.

$$f(z) = \frac{1}{2}z^5 + \frac{2}{3}z^7 + \frac{15}{2}z^9 + \frac{2}{15}z^{11} - \frac{112}{3}z^{13} + 128z^{15} - \frac{512}{3}z^{17} \tag{9.12}$$

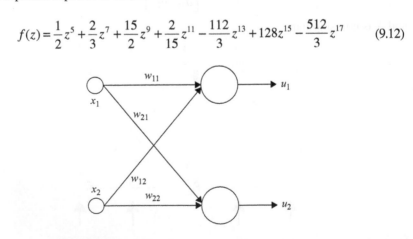

FIGURE 9.5 Single-layered ICA neural network

where z is used to represent the summed input signals being sent to the activation function. The training for such a network is unsupervised using the following weight updating formula:

$$\mathbf{W}(n+1) = W(n) + \eta \left[\mathbf{I} - \mathbf{f}(\mathbf{U}(n))\mathbf{U}^T(n) \right] \mathbf{W}^{-T}(n) \qquad (9.13)$$

Given the fact that the time domain convolution can be converted to multiplication in the frequency domain, frequency domain BBS becomes time domain blind de-convolution.

For example, two microphones at different locations are used to pick up the speech signals as depicted in Figure 9.6.

The received signals by two microphones become

$$\begin{cases} r_1 = s_1 * h_{11} + s_2 * h_{12} \\ r_2 = s_1 * h_{21} + s_2 * h_{22} \end{cases} \qquad (9.14)$$

where s_1 and s_2 are two source signals, r_1 and r_2 are received signals, and $h_{11}, h_{12}, h_{21},$ and h_{22} are impulse responses of related transmission paths.

When short-time Fourier transforms are taken, the above relations become

$$\begin{cases} R_1(\omega,m) = S_1(\omega,m)H_{11}(\omega) + S_2(\omega,m)H_{12}(\omega) \\ R_2(\omega,m) = S_1(\omega,m)H_{21}(\omega) + S_2(\omega,m)H_{22}(\omega) \end{cases} \qquad (9.15)$$

where m is the analysis frame. The short-time Fourier transforms of the speech signals change in different analysis frames, while the Fourier formats of impulse responses remain unchanged (time invariant assumption is applied). In each frequency bin ω, BSS is performed individually. In this way, S_1 and S_2 can be obtained from R_1 and R_2, provided they are statistically independent. This forms the framework of the so-called blind de-convolution (Lee, 1998). Blind de-convolution is in its theoretical infancy and, currently, the separation capability is very limited, especially in dealing with real-world acoustic cases (Mukai et al., 2001; Westner and Bove, 1999).

Since the room impulse responses are involved in the cocktail party problem, simplistic linear mixing and de-mixing using only four coefficients does not work as well as the convoluted ICA model in handing real-world collected signals.

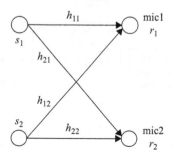

FIGURE 9.6 Two sources and two-microphone system

9.4 BLIND ESTIMATION OF ROOM ACOUSTIC PARAMETERS USING A PCA NETWORK AS A FEATURE EXTRACTOR

As seen in Chapter 8, artificial neural network (ANN) models were developed to accurately extract speech transmission indexes (STIs), reverberation times (RTs), and early decay times (EDTs) from received speech signals. One or one set of ANNs works with a particular speech excitation, as the feature of that particular speech stimulus is built in the ANN through training. This means pre-recorded speech signals have to be used to achieve good accuracy. Feasibility for the above methods to learn from different speech excitations and generalise to arbitrary speech has been explored, but with limited success; they can provide meaningful estimates but fall short for room acoustics measurement applications. This section looks into the use of an additional feature space for speech excitation to achieve better accuracies in blind estimation.

If the modulation transfer function (MTF), as detailed in Chapter 8, can be estimated from the envelope spectra of original and transmitted speech $E_X(F)$ and $E_Y(F)$, STIs and RTs can be extracted subsequently. To achieve blind estimation, information about speech stimulus is needed. One possible way to achieve blind estimation is to implement an addition estimator for the $E_X(F)$ from the received speech. It is found that eigenvalues and eigenvectors of the envelope of received speech signals are correlated with reverberation and features of original speech respectively, providing a useful feature space for this application.

An example is shown in Figure 9.7, where 60-second excerpts of three different running speech signals are convolved with 50 different room impulse responses with RTs from 0.1 to 5 s. Eigenvalues and vectors of the first principal component of envelope signals are observed. Each row is for a speech signal. The left-side plots show the eigenvalues against reverberation times. The right-side ones are over-plots of the eigenvectors for 50 examples with different RTs. In particular, for a given running speech excerpt (45–60 seconds), the principal component eigenvalues of its envelope monotonically decrease when reverberation time increases—see Figure 9.7 (left). Individual speech excitations show distinctive eigenvectors—see Figure 9.7 (right). They are not affected by RTs of the transmission channel. This means that the eigenvectors are related to features of anechoic speech and are robust against the channel characteristics, providing a useful feature space for speech stimuli.

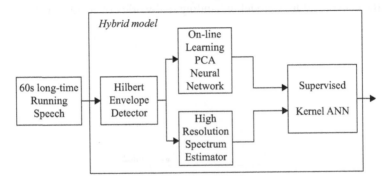

FIGURE 9.7 Eigen values and vectors of first two principle components

Since principal component eigenvectors can be calculated using unsupervised machine learning, this suggests a feasibility of a hybrid ANN model for blind estimation, in which an unsupervised model used to obtain feature space of original speech is added to the supervised models described in previous sections. The hybrid model is illustrated in Figure 9.8.

The envelope detection and envelope spectrum estimation algorithms are identical to the ones discussed in the previous section. But a higher resolution is needed, from immediately above DC to 25 Hz at 0.5 Hz intervals. Eigenvectors of first and second principal components obtained by a principal components analysis (PCA) sub-network from envelope signals and the envelope spectra data are both fed into a supervised neural network. The supervised ANN has distinguishable training and retrieval phases. The PCA subnet is a means to obtain the feature space. So in both training and retrieval phases of the hybrid model, the unsupervised learning to obtain the PCA values is always performed, i.e. it learns "on the fly."

The concept of PCA is not new to statisticians and is often used to identify the most representative low-dimensional subspaces from a high-dimensional vector space. In statistical signal processing and pattern recognition, the PCA offers an effective method for feature selection. It involves finding the eigenvectors and eigenvalues of the correlation matrix of a data set. The PCA network is a neural computing approach to the PCA problem. A speech envelope signal is a time series. It needs to be converted to correlation matrix before PCA can be performed. An m-tap delay-line and a rectangular window are applied to convert speech envelope signals to a multi-dimensional data space for PCA as depicted in Figure 9.9. The speech envelope, low-pass filtered at a cut-off frequency of 15 Hz and decimated to 40 samples/second, is passed through the delay-line and then windowed to obtain m-dimensional observations. A 125–400 ms window is empirically found appropriate. Each column in the data space (reconstruction space) is one observation of the envelope signal. The reconstruction space is used to train a PCA neural network shown in Figure 9.9.

The PCA network has two output neurons and is trained following the generalised Hebbian algorithm describe by Equation 9.6.

The final stage of the hybrid model has a supervised ANN. It takes envelope spectra of received speech and the estimated information about envelopes of original speech signals in terms of eigenvectors as inputs and maps these inputs onto STI or RT values. A typical back-propagation network with two layers of hidden neurons and one linear output neuron as used before is again found adequate. For the standard neuron model having a linear basis function, a sigmoid activation function is used to form the hidden layers. The numbers of neurons on input, two hidden non-linear, and output layers are 60, 50, 35, and 1, respectively.

The training of the hybrid model follows a typical supervised learning regime. Training and validation examples of received speech are obtained via the convolutions of anechoic speech signals with a large number of impulse responses of transmission channels. The trained hybrid network is tested with acoustic and speech cases not seen in the training phase. The maximum prediction error found in the full range tests was 0.087 for STI extraction. Averaging over a few estimates under different speech stimuli gives improved accuracy.

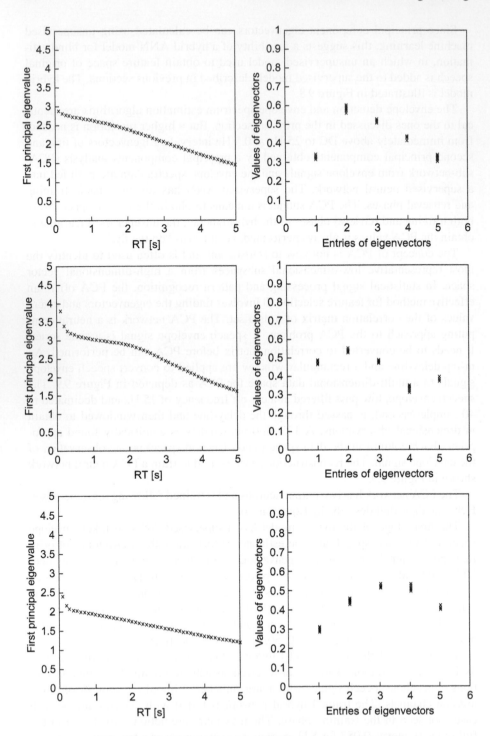

FIGURE 9.8 Hybrid model for blind estimation of STI or RT

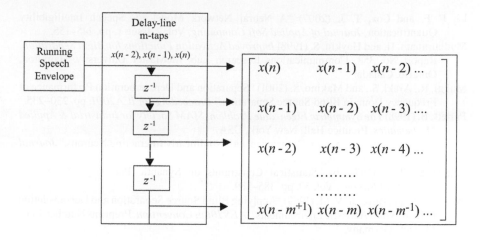

FIGURE 9.9 Converting to correlation matrix format for the PCA network

SUMMARY

Unsupervised learning finds many applications in feature extraction and signal separation. This chapter discussed unsupervised learning and Hebbian learning, and outlined a method to perform PCA through unsupervised learning. ICA, an algorithm closely related to the machine-learning-based PCA, was then presented and discussed in the context of blind source separation and blind de-convolution. It can be demonstrated that blind source separation, or the cocktail party problem as it is known in audio acoustics, may be solved with the ICA algorithm, with some necessary assumptions.

REFERENCES

Becker, S. (1991) "Unsupervised Learning Procedures for Neural Networks," *International Journal of Neural Systems*, Vol. 2, pp. 17–33.

Bell, A. J. and Sejnowski, T. J. (1995) "An Information Maximisation Approach to Blind Separation and Blind De-convolution," *Neural Computation*, Vol. 6, pp. 1129–1159.

Brown, T. H., Kairiss, E. W., and Keenan, C. L. (1990) "Hebbian Synapses: Biophysical Mechanisms and Algorithms," *Annual Review of Neuroscience*, Vol. 13, pp. 475–511.

Changeux, J. P. and Danchin, A. D. (1976) "Selective Stabilisation of Developing Synapses as a Mechanism for the Specification of Neural Networks," *Nature*, Vol. 264, pp. 705–712.

Deco, G. and Obradovic, D. (1996) *An Information-Theoretic Approach to Neural Computing*, Springer-Verlag, New York, USA.

Eggermont, J. J. (1990) "The correlative Brain: *Theory and Experiment in Neural Network*, Springer-Verlag. New York, USA.

Hebb, D. O. (1949) *The Organisation of Behavior: A Neuropsychological Theory*, New York, Wiley.

Lee, T-W. (1998) *Independent Component Analysis—Theory and Applications*, Kluwer Academic Publishers, pp. 83–108.

Li, F. F. and Cox, T. J. (2007) "A Neural Network Model for Speech Intelligibility Quantification, *Journal of Applied Soft Computing*, Vol. 7, Issue 1, pp. 145–155.

Madhuranath, H. and Haykin, S. (1998) *Improved Activation Functions for Blind Separation*, Report No. 358, Communications Research Lab, McMaster University, Hamilton, Ontario, Canada.

Mukai, R., Araki, S., and Makino, S. (2001) "Separation and Dereverberation Performance of Frequency Domain Blind Source Separation," *Proceedings of ICA2001*, pp. 230–235.

Parlett, B. (1998) *The Symmetric Eigenvalue Problem, SIAM Society for Industrial & Applied Mathematics*, Prentice Hall, New York, USA.

Sejnowski, T. J. (1977, a) "Strong Covariance with Nonlinearly Interacting Neurons," *Journal of Mathematical Biology*, Vol. 4, pp. 303–321.

Sejnowski, T. J. (1977, b) "Statistical Constraints on Synaptic Plasticity," *Journal of Theoretical Biology*, Vol. 69, pp. 385–389.

Westner, A. and Bove, Jr. V. M. (1999) "Applying Blind Source Separation and Deconvolution to Real-World Acoustic Environments," *AES 106th Convention*, Preprints Number 4955, Munich, Germany.

BIBLIOGRAPHY AND EXTENDED READING

Haykin, S. (1999) *Neural Networks: A Comprehensive Foundation*, Second Edition, Prentice Hall, Upper Saddle River, New Jersey, USA.

Haykin, S. (2008) *Neural Networks and Learning Machines: A Comprehensive Foundation*, Third Edition, Pearson.

Lee, T-W. (2010) *Independent Component Analysis—Theory and Applications*, Springer, US.

RECOMMENDED SOFTWARE AND TOOL KITS

There is no ICA algorithm in 2016b and earlier versions of MATLAB®. The Statistics and Machine Learning Toolbox on the current version of MATLAB has a function for feature extraction using reconstruction ICA called "rica." This is a function for feature extraction, not straightforward ICA, but the included ICA procedure can be utilised.

http://www.mathworks.com/help/stats/feature-extraction.html#bvmxyf6-1

FastICA package in MATLAB is a popular set of tools broadly used by many researchers. FastICA is also available in C++ and Python.

http://www.cis.hut.fi/projects/ica/fastica/

EXPLORATION AND MINI PROJECT

EXPLORATION

1. Download the *ICA_BSS.zip* and unpack it in the working directory of MATLAB. This zipped file contains heavily commented example/demonstration codes for de-mixing linearly mixed sources.
2. Experiment with different mixtures with a variety of step sizes.

3. Consider a linear mixture generated by Equation 9.10 from two relatively long audio signals, say one minute; write some addition codes so that the de-mixing can be performed segment by segment to mitigate computational overhead. Explore how segment size affects separation.

MINI PROJECT IDEAS

The project ideas are about the separation of two independent audio sources.

- Mix two audio signals (e.g. a chirp and a speech signal, or female and male speech signals) by an arbitrary mixing matrix according to Equation 10. This is to emulate two independent sources picked up by two microphones at unequal distances from sources in an acoustically dry space or location free from major reflective surfaces. Use the ICA blind source separation to separate these two sources. With reference to the result, give critiques of the method for blind source separation in the context of real-world application of the ICA for audio source separation.

- It is known that ICA BSS algorithms do not differentiate which source is which. Develop a simple method to differentiate the sources using up to three audio features, e.g., zero crossing rate, crest factor, entropy, ketosis, spectral centroids, depending on the types of signals for separation.

- Consider the solution to the same cocktail party problem but in a reverberant space. The signals are, therefore, mixed according to Equation 9.17. Explore how time domain ICA behaves on these mixtures, develop a frequency domain ICA algorithm following the idea as outlined in Equation 9.18, and compare its performance with that of the time domain ICA. Note that real room impulse responses represent mixed phase systems. They are complicated and very long. As a mini project, much simplified emulation of room impulse responses may be considered. Four low pass FIR filters of similar frequency response (say around 3 dB/Oct rolling off from 250 Hz) but designed by different approximation methods or having different number of taps may be used to represent to four impulse responses in Equation 9.17.

10 DSP in Hearing Aids

Hearing aids are used to help people with mitigated hearing, including congenital, acquired, and age-related hearing losses or impairments. With the advancement of new technologies, especially microelectronics enabled "system on chip," hearing aids can be made more cosmetically acceptable, functionally practical, comfortable, and easier to use. Nowadays, they are used more and more to improve the quality of life of many people.

The earliest hearing aids may be traced back to the 17th century. The instruments were essentially passive acoustic devices taking certain forms of reversed horns known as ear trumpets or ear horns. There were either ad hoc or improvised ways to help with hearing problems. For example, Ludwig von Beethoven used a metal rod that he could bite at one end; he attached the other end to the sound board of his piano. In this way, he could use bone-conducted vibrations to help hear what he played from the piano.

Hearing aids have been developed and have evolved with the advancement of electroacoustics. In the earliest electronic hearing aids developed in the late 1800s, carbon transmitters were used so that acoustic signals were amplified by the current without resorting to amplifiers. These hearing aids were literally the same as the early telephones, but with the sound level at ear pieces lifted. The first generation of electronically amplified hearing aids using vacuum tube amplifiers were developed and made commercially available from 1910 to 1920. Further development towards miniaturisation of vacuum tube based hearing aids was seen from 1920 to 1950. The invention of transistors in 1948 advanced the miniaturisation of electronics, including hearing aids. The development of digital signal processing techniques and emerging of micro-processors in the 1970s started a completely new era of digital hearing aids. Technologies such as adaptive feedback cancellation, multi-channel compression, noise suppression, and frequency shifting have been introduced to hearing aid applications. Hearing aid systems on chips were developed in the 1980s.

Modern hearing aids typically have three major building blocks, namely sound acquisition by a microphone or more microphones, signal processing and amplification with DSP techniques, and sound reproduction using high-quality transducers at suitable levels. It is evident that the ever-increasing use of DSP represents the trend of advancement of hearing aids to address various challenges in this field. It is not imprudent to assert that DSP is the key and enabling technology for modern hearing aids. On the other hand, many audio and acoustic DSP techniques find flourish paradigms in hearing aids.

The fundamental aspects and techniques of DSP used in hearing aids are nothing special. They include amplification, filtering, equalisation, noise reduction, feedback cancellation, speech enhancement noise classification, and frequency shifting. Many of these have been introduced and discussed in previous chapters. Nevertheless, there are a good number of challenges, constraints, and special considerations for these techniques in hearing aid applications. This chapter will highlight these special aspects.

10.1   TECHNICAL CHALLENGES OF HEARING AIDS

As wearable appliances, miniaturisation, cosmetic appearance, battery life, and ease of use are all important concerns of hearing aids, and sometimes these concerns become constraints. As part of a DSP book, the focus of this chapter will be the challenges imposed on signal processing algorithms and typical DSP building blocks found in modern hearing aids.

Hearing losses are frequency related, i.e. one may have mitigated hearing in certain frequency bands but not across all frequencies. Equal amplification of acoustic signals across all audible frequencies can make the users feel it is too loud or noisy to wear these devices. Normal human hearing has a specific dynamic range from audible threshold to threshold of pain. Hearing losses shift the audible threshold upwards, resulting in a reduced dynamic range. Straightforward amplification has limited scope to satisfactory user experience and, therefore, suitable compressions are sought. Modern hearing aids typically integrate microphones and a speaker in a single wearable housing. The proximity of microphones and speakers makes them prone to acoustic feedback problems. Suitable feedback cancellation is often used. Ambient noises are broadband in nature; they become a nuisance if excessive background noise is amplified and perceived. Noise reduction represents another useful algorithm for hearing aids. The loss of speech intelligibility is often the reason why users start to wear hearing aids. For speech signals to be intelligible, certain critical frequency components in the signal spectra must be sensed by listeners. If one has severe losses of hearing in these frequencies, straightforward amplification and dynamic range compression can hardly achieve satisfactory improvements. In such cases, frequency transposition, typically frequency lowering, can be considered, i.e. shifting the frequencies of the sound of interest to a frequency range in which the users can effectively sense the signals. Other DSP blocks in high-end hearing aids include directional information processing, which helps the user regain some source localisation ability, and environmental noise classification algorithms, which enable the device to adapt to the ambient environment better. Modern hearing aids can have a number of function blocks, as illustrated in Figure 10.1. ANSI sub-band number 22 (150 Hz) to number 39 (7.94 kHz) are individually amplified with desirable gains and merged into a number of channels for compression. So in hearing aids, band numbers and channel numbers are different concepts; they should not be confused.

10.2   AUDIOMETRY AND HEARING AID FITTING

In audiology, audiometry is used to assess human hearing acuity and diagnose hearing losses. The audiometry procedure is typically undertaken using an audiometer. The result of the examination is plotted in a graph called an audiogram. The audiogram has a logarithmic frequency scale as its horizontal axis and a linear (in dB) vertical axis. Figure 10.2 shows an audiogram sheet. The vertical axis reading is dB hearing level (dBHL). A basic audiogram starts at a low frequency of 125 Hz and increases at octave intervals to a high frequency of 8 kHz. The dBHL is defined for audiometry with pure tone, and should be differentiated from other dB scales used in electronics and acoustics.

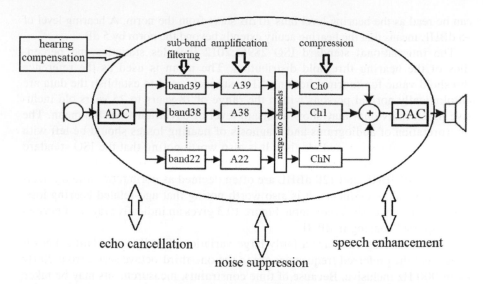

FIGURE 10.1   A block diagram illustration of modern hearing aids

Normal human hearing shows different sensitivities to each frequency band across the audible spectrum. A norm of threshold of hearing at each frequency band, as established in the ISO standard (ISO, 2012) is indicated by 0 dBHL. That is to say, 0 dBHL represents the threshold of hearing of pure tones of each frequency for a normal-hearing population. A hearing level of +10 dBHL in a particular frequency band means a 10 dB compensation is needed to achieve normal hearing in that frequency band, or this

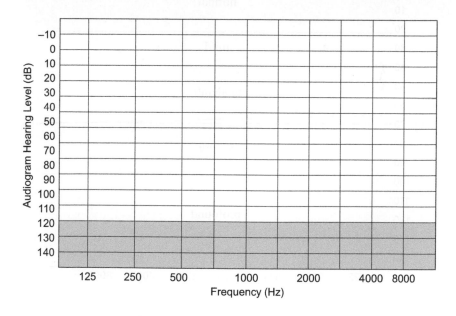

FIGURE 10.2   An audiogram sheet

can be read as the hearing level falls 10 dB short from the norm. A hearing level of −5 dBHL means that the hearing acuity extends beyond the norm by 5 dB.

The international standard ISO 28961:2012 includes statistics and percentiles of the hearing threshold distribution. The mean is used as the reference threshold value for "normal hearing." The subjects used to establish the data are ontologically normal persons within the range of 18 years to 25 years old inclusive. Each individual may show a certain level of deviation from the mean. The interpretation of audiograms and diagnosis of hearing losses should be left with trained specialists e.g. audiologists. It is also worth noting that the ISO standard is based on the statistics of a younger population. For an individual, hearing levels between −10 dBHL and +20 dBHL are often deemed as normal or ordinary hearing in clinical assessments. It is also worth noting that age-related hearing loss, known as presbycusis, is common. Figure 10.3 gives an indicative range of normal and impaired hearing in dBHL.

Normal hearing can have a fairly large variation. The ISO standard acknowledges that the preferred frequencies are in the one-third-octave series from 20 Hz to 16 000 Hz inclusive. Because of time constraints, measurements may be taken only at 125, 250, 500, 1k, 2k, 3k, 4k, 6k, and 8k Hz and two (left and right) audiograms may be recorded on the same chart by changing the colour of the pens. Table 10.1 shows 0 dBHL and equivalent dB for sound level pressure (SPL) across octave bands.

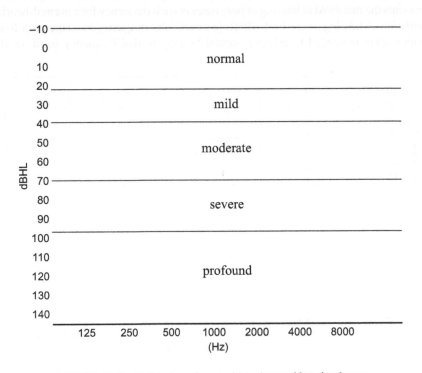

FIGURE 10.3 Indication of normal hearing and hearing losses

TABLE 10.1

0 dBHL and Equivalent dB SPL

Frequency (Hz)	250	500	1k	2k	3k	4k	6k	8k
Equivalent Threshold L_p (dB)	25.5	11.5	7.0	9.0	10.0	9.5	15.5	13.0

There have been around a dozen hearing aids fitting protocols in the past along with the evolution of hearing aids. They developed, evolved, and some of them were phased out eventually. These methods were essentially developed to achieve optimal results with the very limited number of adjustable parameters offered by the then-available technologies. Nowadays, with high-end and modern digital hearing aids, audiologists can almost fit to prescriptive targets for most patients. The procedures are converged to a few common ones. The NAL (National Acoustics Laboratories, Australia) methods are probably the most popular amongst them.

The NAL family fitting protocols, including NAL, NAL-R, NAL-NL1, and NAL-NL2, attempt to equalise signals across speech frequencies to achieve optimal results for individual patients. The original version of the NAL fitting procedure was developed in 1976 (Byrne & Tonnison, 1976). User feedback and research showed this version of the NAL prescribed too low a low-frequency gain and too high a high-frequency gain for patients with significant high-frequency hearing losses. To address this issue, the revised version known as NAL-R was developed by Byrne & Dillon (1986). Further development led to the NAL-RP (Byrne et al., 1990), which extended the NAL-R to include specific consideration for severe hearing loss cases.

The goal of hearing aids is not simply to compensate for the hearing losses to get close to 0 dBHL hearing. Instead, it aims to provide users with maximum speech intelligibility or "effective audibility." Psychoacoustic models were considered, leading to the non-linear hearing aids fitting (Dillon et al., 1998; Byrne et al., 2001) known as the NAL-NL1 procedure. The objectives of NAL-NL1 fitting are to maximise speech intelligibility and to limit the loudness to a normal or even lower level. The criteria used to determine intelligibility are taken from the Speech Intelligibility Index (SII) (ANSI S3.5, 1997), which is a revision of the older version ANSI S3.5, 1969, "American National Standard Methods for the Calculation of the Articulation Index."

The newest fitting procedure from the NAL family is the version of the NAL known as NAL-NL2. The goals of the NAL-NL2 are the same as those for the NAL-NL1, i.e. to maximise speech intelligibility but keep the loudness less than or equal to normal loudness. Nonetheless, the NAL-NL2 reflects improvements made based on the feedback of ten years' practice with the NAL-NL1 and continuous research. There is an edited transcript of a live seminar given by the developers of the NAL-NL2 procedure on Audiology Online, which is a useful document for the understanding of the new version from the NAL family (Keidser et al., 2011).

The procedures are made easier for audiologists by software packages. The actual procedure is to acquire audiograms for patients first and input the data into the NAL software package, which calculates all programming parameters for digital hearing aids fitting. Of course, audiologists often modify these parameters based on many other considerations.

10.3 FILTER BANK AND MULTI-BAND COMPRESSION

Modern digital hearing aids can have a number of function blocks, as was illustrated in Figure 10.1. The auditory compensation is the essential functional block that all hearing aids must have. From a signal processing perspective, it is a programmable multi-frequency sub-band equaliser and compressor, which allows for flexible manipulation of amplification gains and compression ratios in individual frequency bands by audiologists to achieve the goals of their prescription.

10.3.1 FILTER BANK

Amplification gain adjustment in individual frequency bands is achieved by decomposing the full audio signals into different bands by a number of band pass filters taking the form of a filter bank. Band passed signals are gain adjusted and fed into the subsequent compression channels. The number of bands affects an audiologist's ability to modify the frequency responses to achieve the best fit for individual users to best restore their hearing. On the other hand, more sub-bands can often help better control acoustic feedback without compromising fitting accuracy.

The DFT and its fast algorithm FFT prevalently used in DSP are intrinsically filter banks. FFT-based filter banks do find their applications in hearing aids. However, FFT-based filter bands are naturally uniformly spaced in frequency, which defers from the non-uniformly specified frequency bands in audiograms and hearing and prescriptions for hearing aids. The psychoacoustic model underpinning hearing aids fitting is the Articulation Index and/or Speech Intelligibility Index. Articulation Index (AI) is calculated from the 1/3 octave band levels between 200 Hz and 6300 Hz centre frequencies. Each of the 1/3 octave dB(A) levels are weighted according to the following protocols:

- If A-weighted 1/3 octave level lies between upper and lower limits, then it takes a linear value between 0 and 1.
- If value falls above the upper limit, then result = 0 for that particular 1/3 octave band.
- If value falls below the lower limit, then result = 1 for that particular 1/3 octave band.
- Multiply all of the calculated values by the AI weighting and sum all the values to get the AI in percentage terms.

The upper and lower limits and weighting factor for AI calculations are given in Table 10.2.

In addition to audiometry, many other auditory perception models are based on 1/3 octave band descriptions. This leads to the standardisation of the characteristics of 1/3 filter banks known as ANSI S11.11 (ANSI, 2004). It is the counterpart and generally confirms the specification of IEC 61260:1995. In many other audio applications, centre frequencies are often quoted to specify a particular sub-band.

TABLE 10.2

Upper and Lower Limits and Sub-band Weighting

1/3 Octave Centre Frequency (Hz)	AI Lower Level dB(A)	AI Upper Level dB(A)	AI Weighting
200	23.1	53.1	1.0
250	30.4	60.4	2.0
315	34.4	64.4	3.25
400	38.2	68.2	4.25
500	41.8	71.8	4.5
630	43.1	73.1	5.25
800	44.2	74.2	6.5
1000	44.0	74.0	7.25
1250	42.6	72.6	8.5
1600	41.0	71.0	11.5
2000	38.2	68.2	11
2500	36.3	66.3	9.5
3150	34.2	64.2	9.0
4000	31.0	61.0	7.75
5000	26.5	56.5	6.25
6300	20.9	50.9	2.5

The ANSI standard uses band numbers to name the 1/3 octave bands but centre frequencies for octave bands. Figure 10.1 uses band numbers to illustrate the filter bank. Audio bands start from ANSI Band 14 to Band 43, including 30 1/3-octave bands or 10 octave bands.

A prevalent and accurate way for the assessment of hearing losses and prescribed insertion gain follows the 1/3-octave frequency bands. The design of octave-band filter banks and fractional octave band filter banks are essentially the synthesis of a series of band pass filters to specifications. Given that the ANSI S1.11 details the design specification, this is often followed in various audio-related applications, especially in measurements so that the accuracy can be easily quoted with a standard as reference. The ANSI S1.11-2004 standard defines three classes in terms of ripples in pass bands and attenuation levels in stop band: class 0, class 1, and class 2 for +/−0.15 dB, +/−0.3 dB, and +/−0.5 dB, respectively. Levels of stop band attenuation range from 60 to 75dB depending on the class of the filter. The design of octave and fractional octave filter banks is given as a function in MATLAB®: fdesign.octave(BandsPerOctave,'Class 2','N,F0',N,F0,Fs), where N is filter order, F0 is centre frequency, and Fs the sampling frequency. This is quite straightforward if there is no stipulated phase specification and other constraints. It is worth noting that, in addition to keeping pass band ripples low and stop band attention high, one key issue for octave and fractional octave band filter bank design is to ensure that the sum of file bank output gives a flat spectrum, i.e. if a white noise is passed through the filter bank, the sum of all outputs should remain

TABLE 10.3

ANSI Octave and 1/3 Octave Bands in Audio Frequencies

Band number x	Base-ten exact f_m $(10^{x/10})$, Hz	Base-two exact f_m $(2^{(x-30)/3})(1000)$, Hz	Nominal midband frequency, Hz	Octave
14	25.119	24.803	25	*
15	31.623	31.250	31.5	
16	39.811	39.373	40	
17	50.119	49.606	50	*
18	63.096	62.500	63	
19	79.433	78.745	80	
20	100.00	99.213	100	*
21	125.89	125.00	125	
22	158.49	157.49	160	
23	199.53	198.43	200	*
24	251.19	250.00	250	
25	316.23	314.98	315	
26	398.11	396.85	400	*
27	501.19	500.00	500	
28	630.96	629.96	630	
29	794.33	793.70	800	*
30	1,000.0	1,000.0	1 000	
31	1,258.9	1,259.9	1 250	
32	1,584.9	1,587.4	1 600	*
33	1,995.3	2,000.0	2 000	
34	2,511.9	2,519.8	2 500	
35	3,162.3	3,174.8	3 150	*
36	3,981.1	4,000.0	4 000	
37	5,011.9	5,039.7	5 000	
38	6,309.6	6,349.6	6 300	*
39	7,943.3	8,000.0	8 000	
40	10,000	10,079	10 000	
41	12,589	12,699	12 500	*
42	15,849	16,000	16 000	
43	19,953	20,159	20 000	

to be a white noise. Figure 10.4 illustrates a 1/3 octave band filter implemented using MATLAB.

The ANSI and IEC standard octave and fractional octave band filter banks have been defined for a variety of uses, including instrumentation. They call for rather high precision as shown in Table 10.3. Recall the dBHLs for normal hearing indicated in Figure 10.3; they show a large swing. This means the sort of accuracy specified in the ANSI standard is rarely needed for hearing aids applications. Quasi ANSI standard filter banks with lowered accuracy are often used for hearing aids to mitigate other design constrains. For non-real-time simulation and sample generation,

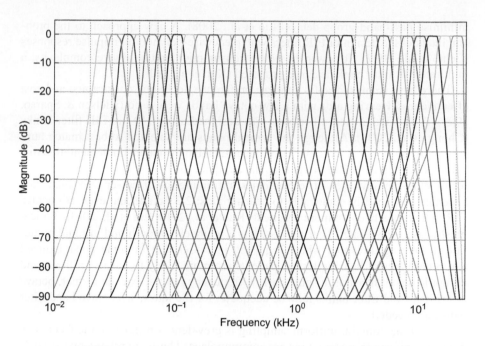

FIGURE 10.4 A MATLAB implementation of 1/3-octave band filter

the filter bank design tools available from shelfware, such as MATLAB, are more than adequate if phase responses and hardware constraints are not specific concerns.

Most ANSI S1.11 filter banks found in the literature are implemented with infinite impulse response (IIR) structures, taking advantage of higher efficiency in hardware usage, and lower computational and structural complexity. Indeed, normal human hearing is not particularly sensitive to phase-distortion in monaural settings. However, FIR filter banks are preferred in hearing aids and adopted for their linear phase, stability, and simple structures. Basic hearing aids are design mainly for effective intelligibility. More advanced devices take into account music listening and source localisation. Phase relationships are found important in music perception, particularly for tones of lower frequencies with rich harmonic content. Linear phase FIR filter bands are the choice for hearing aids taking into account music listening (Chasm & Russo, 2004). Furthermore, the linear-phase property makes the processed sound more natural for other sources. It is well understood that interaural time delays are important cues for source localisation. Linear phase FIR filter banks may help preserve phase cues and, in turn, source localisation ability. Moreover, the use of linear phase FIR filter banks arguably makes feedback cancellation easier (Chong et al., 2006).

FIR filter banks are known to be hardware resource and power consuming. In implementation of algorithms for hearing aids, there are several constraints from the hardware platforms, as the miniaturisation, power consumption, and real-time requirements all limit the actual implementation of the computationally intensive algorithms. As a result, there are continuous efforts in the design of filter bands to minimise pass-band group delays (Liu et al., 2008). A new approach to the problem includes the design of FIR filters with piecewise-polynomial impulse responses (Lehto et al., 2007). To lower the power consumption and hardware complexity, a multirate FIR filter bank design is proposed (Kuo et al., 2010).

There have been trade-offs amongst preference, technical constraints, and cost effectiveness. Hearing aids with a few octave bands were used (Nielsen & Sparso, 1999). A 16-band critical filter bank was designed using 110-tap FIR filter banks (Chong et al., 2006). Although the critical bands are selected to best match hum speech perception, the irregularity of the critical bands makes implementation difficult. A 110-tap FIR filter is a non-trivial computation load- on a hearing aids platform. With the advances of technologies, miniature lower-power DSP platforms become more readily deployable to hearing aids; 2/3 octave and even 1/3 octave FIR filter banks become more practical (Kuo et al., 2010; Liu et al., 2013). Typical presbycusis (age-related hearing loss) has a relatively smooth and gradual reduction in hearing threshold when frequency increases, but in some noise-induced and other types of hearing losses, patients are more likely to have significant and abrupt losses of hearing in some narrow frequency ranges. For the former, octave or critical band filter banks may be adequate; for the latter, 1/3 octave band filter banks are needed.

In the time domain, uniform sampling is prevalent; similarly in the frequency domain, uniform frequency bins are commonplace. These are related by the DFT. The availability of FFT algorithms makes uniform sampling a predominant way to present signals in the time and frequency domains. A straightforward advantage of using this uniformity in the frequency domain is that filter bands can be designed directly using DFT, e.g. a 32-band FFT filter bank in digital hearing aids (Brennan & Schneider, 1998). However, such uniform frequency sampling does not represent human perception in audio signals. This gives rise to the question of whether the filter bank should be designed in the first instance as uniform filter banks and then convert them to non-uniform 1/3 octave bands, or start the design at the very beginning with a non-uniform format. Both approaches have been used in the design of filter banks for hearing aids. Each has its own advantages and disadvantages. Obviously, uniform filter banks are easier to design, since it is in the native format of frequency domain. But a uniform filter bank requires an additional conversion stage that maps the linearly specified sub-bands onto the non-uniform frequency format to match the hearing aids prescriptions. Attempts were made to implement computationally efficiency uniform filter banks (Lian & Wei, 2005).

10.3.2 Compression Channel

Hearing loss is because of the elevated hearing threshold. However, this does not mean an equal elevation of the ceiling levels. In fact, the ceiling level normally remains unchanged. The reduced dynamic range means that a linear amplifier that can get the weak sound above hearing threshold can make the loud sound too loud for the users. Audio compression and limiting aim to reduce the dynamic range of the audio signals and have been used in recording, broadcasting, and many other audio applications; they have also found applications in hearing aids to address the dynamic range issue.

Like any other audio compressors, compression schemes in hearing aids have four major parameters, namely threshold, compression ratio, and attach and release times. Channel compression in hearing aids starts from defining a compression threshold. The compression takes place when the signal level is above this threshold. The compression is a straightforward non-linear mapping relation between input and output. For a compression scheme with a compression ratio of c switched in (when the input level becomes higher than the threshold), a non-linear relationship between output and input is established. Compression ratio c defines such non-linear relationships. Compression ratio is the change in input level in dB required to achieve 1 dB change in the output level; in other words, 1 dB increase in the input signal results in a $1/c$ dB output increase. Figure 10.5 shows the input-output relationships of different compression ratios. At the threshold, there is an abrupt change; it can make the users feel a sudden or unsmooth change. To address this problem, a soft threshold transition or "knee" can be introduced.

Since the residual dynamic ranges, i.e. from hearing threshold to ceiling level, in each sub-band for individual patients are different, the compression ratios differ in sub-bands and need to be adjusted individually. The compressions applied to individual sub-bands (often denoted by ANSI standard band numbers) form the compression channels (Figure 10.1).

To avoid sudden engaging and disengaging of the compression, slow transitions are introduced and time-constants, namely attach time and release time (also known as recovery time), are used to specify the "speed," as illustrated in Figure 10.6. Attach time is usually shorter than release time. As ballpark figures, attach time is usually around 5 ms and release time around 25 ms. The compression speed adjustment is also important for hearing aids fitting to achieve the goal and maintain

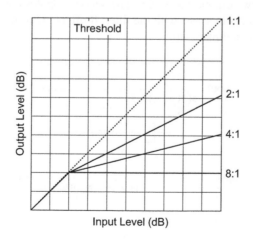

FIGURE 10.5   Input-output relationships vs compression ratios

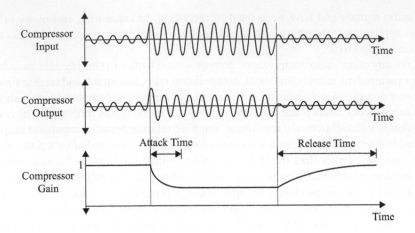

FIGURE 10.6   Attach and release time

comfort in use. There are many practical considerations from audiologists (Brennan & Schneider, 1998). When attack and release time becomes very long, say several hundred milliseconds, the compression literally becomes automatic gain control (AGC) used in some old single-channel hearing aids.

Two types of compression systems as illustrated in Figure 10.7, namely feed forward and feedback, are commonly used. A feedback compression system uses the filtered output signals to control the system gain. Since the feedback uses error to reduce error, so error always exists; otherwise, there would be no control. In compressor design, the use of feedback control mechanisms means that overshot is inevitable. For analogue circuit design, feedback topologies are relative simple. Feedback topologies were commonly seen in old-type signal channel hearing aids, sometimes implemented as slow responding AGCs. The second topology is the feed forward one. The gain control is in series with the filter. It offers great flexibility and is easy to implement on a digital platform.

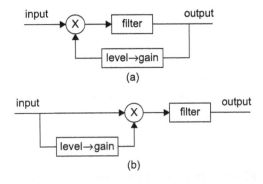

FIGURE 10.7   Feed forward and feedback topologies

10.4 ACOUSTIC FEEDBACK CANCELLATION

An earpiece with a long cable connected to a portable radio like box was the image of hearing aids in the past. Modern hearing aids, such as in the canal, in the ear, and behind the ear models, are much smaller and often a one-piece design. This means that the distance between microphone(s) and the speaker are much closer. Acoustics feedback, if not suitably cancelled, will limit the required amplification gains. Acoustics feedback in hearing aids refers to the fact that part of the acoustic signals produced by the speaker are leaked into the air or transmitted as structure-borne sound and fed backwards to the microphone(s). Acoustic feedback due to airborne sound is mainly because of the leakage from the ear canal through a vent or loose fitting, while structure-borne acoustic feedback mainly results from the housing of the device.

A simple feedback loop illustrated in Figure 10.8 can be used to explain the stability issue caused by acoustic feedback, in which G is the forward path transfer function, i.e. the intended frequency dependent amplification and compression; H is the lump model of the transfer function of all acoustic feedbacks.

The parasitic feedback loop makes the hearing aids a feedback system. The close-loop gain G' becomes

$$G' = G/(1 - GH) \tag{10.1}$$

It is apparent, when loop gain GH is greater than 1 and feedback is positive, i.e. $|GH| > 1$ and $\arg(GH) = 2\pi n$ rad, where n is an integer, the feedback loop becomes unstable and turns itself into an oscillator. The problem of acoustic feedback caused oscillation is known by the users as "whistling" or "howling," depending upon the frequencies. It is extremely annoying to the users. The conditions that can trigger hearing aids to oscillate are prevailing. Typically, the feedback path H can attenuate acoustic signals by 20 to 50 dB, while the intended amplification is usually 15 to 50 dB, depending on frequency bands (Liu et al., 2013). In theory, by keeping away from any one of the aforementioned phase and magnitude conditions, oscillation can be eliminated. Nonetheless, in hearing aids, the feedback path changes significantly with frequencies. Within the frequency range of interest, multiple 360 degrees phase changes take place. It is, therefore, sensible to focus on the loop gain only. In many cases, the allowable amplification is simply too small to compensate for the hearing loss. Therefore, feedback suppression algorithms are necessary and commonly used in modern digital hearing aids, and automatic tuning notch filters or adaptive filters are often used to cancel the feedback induced oscillation.

When oscillation occurs, there is a dominant frequency, depending upon the loop transfer function. One possible way to suppress the acoustic feedback is to insert a

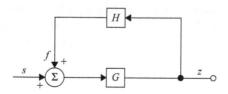

FIGURE 10.8 Acoustic feedback loop

notch filter in cascade with the forward path with its centre frequency tuned to the peak found in the signal spectrum, since the acoustic feedback path is deemed as time variant. This is quite true, as the acoustic transmission path can be changed by surrounding reflective or absorptive surfaces when the users move around. It is also affected by the fitting. Furthermore, the loop gain changes with the input signal due to the non-linear compression.

A second-order adaptive notch filter was sucessfully implemented by Maxwell & Zurek (1995) based on the filter proposed via simulation by Kates (1991). The difference equations for the filter are

$$u[n] = t[n] + \rho a[n-1]u[n-1] - \rho^2 u[n-2] \tag{10.2}$$

$$u[n] = u[n] - a[n-1]u[n-1] + u[n-2] \tag{10.3}$$

where $t[n]$ is the input of the filter, $u[n]$ is the output of the filter, ρ is the pole radius, $a[n]$ is the notch filter parameter following the LMS update formula

$$a[n] = a[n-1] + 2\mu u[n-1] + u[n-2] \tag{10.4}$$

μ is adaptation time constant and the running reduction by the notch filter is estimated

$$p[n] = \beta p[n-1] + (1-\beta)\left(\|t[n]\| - |v[n]|\right) \tag{10.5}$$

$a[n]$ was constrained to be less than 2 to maintain stability, μ was chosen to be 10^{-12} and β 0.99.

Implementation and evaluation suggest that the algorithm works well for frequencies above 500 Hz, but in lower frequencies, the algorithm is prone to misdetect significant spectral peak in vowels as oscillation.

The notch filter outlined above for acoustic feedback cancellation may be viewed and is, in fact, named after adaptive notch filtering because there is an adaptive/ updating algorithm to track the peak in the spectrum and modify the centre frequency of the filter. Nonetheless, it is not a strict sense adaptive filter, in the sense that it does not estimate and minimise the error. These sorts of true adaptive filters have become more commonly used in hearing aids for acoustic feedback cancellation in recent years. Figure 10.9 shows a typical true adaptive scheme for acoustic feedback cancellation.

The inner block in Figure 10.9 is the essential auditory compensation block for hearing aids, but with an adaptive filter added. The outer loop is the parasitic acoustic feedback path. $s[n]$ is the source (speech) signal, n is the sample number, and $y[n]$ is the output signal. If no adaptive filter is added, the z-transform function of the system taking into account the acoustic feedback can be re-written according to Equation 10.1 as

$$A(z) = \frac{z^{-D} \cdot G(2)}{1 - z^{-D}G(z) \cdot H(z)} \tag{10.6}$$

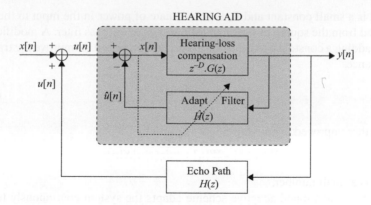

FIGURE 10.9 A true adaptive scheme for acoustic feedback cancellation

The added feedback cancellation path $H'(z)$ is an LMS adaptive filter. It attempts to estimate the acsoutic feedback, so that the estimated feedback can be taken out from the microphone signal $v(n)$. The estimation is achieved by minimising the error signal $x[n]$ in a least mean square sense.

$$Min \ x[n] = Min(v[n] - v'[n]) \qquad (10.7)$$

The close loop transfer function when the adaptive acoustic feedback cancellation path is added becomes

$$A(z) = \frac{z^{-D} \cdot G(2)}{1 - z^{-D} G(z) \cdot \left(H(z) - H'^{(z)} \right)} \qquad (10.8)$$

For system stability, the constraint is

$$\left| G(z) \cdot \left(H(z) - H'^{(z)} \right) \right| < 1 \qquad (10.9)$$

The LMS algorithms used to find the filter coefficients are similar to those discussed before, such as the ones used to reduce noise. Some special tailored versions were proposed to improve the performance in hearing aids applications. A variety of modified LMS algorithms have been developed for the updating of the filter coefficient vector **W** for an L-tap filter; a detailed review and evaluation of their performances were carried out via simulation (Maxwell & Zurek, 1995). The standard LMS updating formulae are

$$\varepsilon[n] = s[n] - W^T[n]X[n] \qquad (10.10)$$

where $X[n]$ is the vector of L-point data and

$$W[n + 1] = W[n] + 2\mu\varepsilon[n]X[n] \qquad (10.11)$$

$$\mu = \frac{\alpha}{LP_x} \qquad (10.12)$$

where α is a small constant and P_x is an estimate of power in the input to the filter, calculated from the square of the input followed by a low pass filter. A modified version, by adding a constant c to Equation 10.12, was proposed to improve the transient performance.

$$\mu = \frac{\alpha}{LPx + c} \tag{10.13}$$

A further improved update formula was given by

$$W[n+1] = W[n] + ksig(\varepsilon[n]X[n]) \tag{10.14}$$

where k is a small number.

The above mentioned adaptive scheme adapts the system continuously to minimise the error, using the estimated acoustic feedback. There is another type of adaptive filter technique being used in hearing aids acoustic feedback cancellation, known as non-continuous adaptation acoustic feedback cancellation. In such a system, the acoustic feedback path is "measured" using a white noise or other suitable test stimulus. The measured acoustic feedback path transfer function is used to determine the error that is to be deducted from the microphone signals. Since the measurement is taken at the very beginning and at certain intervals, hence the name non-continuous adaptation. When compared with continuous counterparts, the non-continuous ones can more precisely determine the acoustic feedback channel; however, regularly occurring probe stimuli are certainly not welcome by users. In between two measurements, such systems do not adapt themselves to environment changes.

10.5   TRANSPOSITION AND FREQUENCY LOWERING

For patients with severe hearing losses in a particular frequency band or frequency bands, typically in high frequencies, hearing aids may not be able to reach the required amplification gains in these frequency bands before acoustic feedback occurs. In other cases, the hearing loss in certain frequency bands is so severe that it cannot be compensated for by amplification. This is known as "dead region(s) in cochlea" (Moore, 2004). If these dead regions happen in critical bands for speech, simple amplification will not help with effective intelligibility. One solution for such cases is the so-called frequency transposing, which shifts the frequency components in the dead region(s) to the region(s) where hearing is not so badly impaired. In clinical settings, the required transposition for most patients is to transpose high frequency components in a lower frequency band or bands' hence, frequency lowering is the common and probably the earlier transposition method in hearing aids applications. Although transposing hearing aids may sound unnatural, after proper training, users can gradually get to regain some of the "hearing" in the frequency bands they cannot hear. Frequency lowering can be achieved quite easily using the "slow playback," i.e. to oversample the signal's high frequency bands and then play back at normal sampling rate. This earlier version of frequency lowering often lowers the frequency to a lower octave band. VOCODER frequency lowering is another

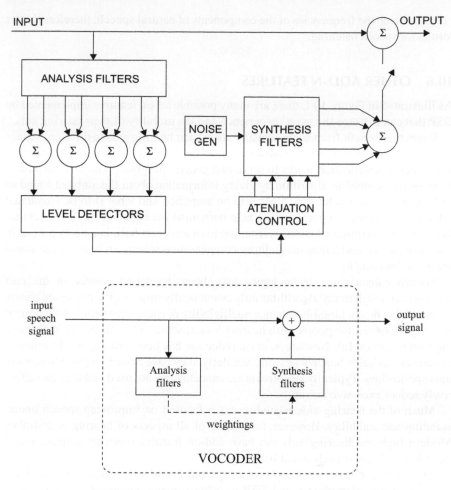

FIGURE 10.10   VOCODER frequency lowering (a, b)

possible way to achieve the design goal of frequency transposition in hearing aids. Figure 10.10 illustrates a block diagram of a VOCODER frequency lowering scheme (Posen et al., 1993).

High frequency components in speech signals from 1 to 5 kHz are analysed using a bank of 8 1/3 octave band filters. Two adjacent bands were merged to give 4 bands for analysis. Typical 20 ms frames were considered and the output levels of these bands calculated and subsequently used to control the levels in the synthesis filter. A broadband noise generator is used. The noise is filtered by 4 1/3 octave band synthesis filters from 394 to 793 Hz, the output levels of the synthesis filters are controlled by the levels of detected speech signals so that a linear proportion relation is retained. Apparently, this scheme transposed speech components in 1 to 5 kHz down to around 400 to 800 Hz.

A more advanced frequency lowering can be achieved using the proportional frequency compression technique (Turner & Hurtig, 1999). This method preserves the ratios between the frequencies of the components of natural speech; therefore, better results for speech listening.

10.6   OTHER ADD-N FEATURES

As illustrated in Figure 10.1, there are many possible ad-on features implemented by DSP that can enhance the users' experience and the usability of their hearing aids.

Users may benefit from some adjustments to their hearing aids in different venues and for different type of sound sources. Sound classification is introduced to hearing aids. Sound classification works to serve two general purposes. The first purpose is to assist the de-noising algorithm by giving information about the ambient sound so that the system can determine what is to be amplified and what is to be attenuated selectively. The second purpose is to help determine the equalisation. Sound classification can determine what one is listening to in a concert hall, in a noisy pub, or in a lecture theatre, and adjust the auditory compensation scheme to best suit the sound users are listening to.

Speech enhancement technologies have been developed rapidly in the past 20 years, although many algorithms only cosmetically improve the perceived clarity of speech but do not tangibly enhance intelligibility of speech for people with normal hearing. However, for people with hearing losses, some of the speech enhancement algorithms are useful. Nowadays, noise reduction has been broadly used in mobile phones and aviation headphones, etc. Similarly, they have been brought into hearing aids applications. Typically, adaptive noise cancellations are used and they can effectively reduce excessive noises.

Much of the hearing aids related research focused on improving speech understanding and audibility. However, restoration of all aspects of hearing is desirable. Modern high-end hearing aids can have add-on features that can improve users' experience of non-speech sound listening.

Multi-microphones or microphone arrays and beamforming use the directional patterns of the microphones and DSP to achieve many advanced features. These new technologies can improve noise reduction, sound classification, and sound localisation.

SUMMARY

Modern and advanced hearing aids are a typical example of integrated implementation of a variety of audio DSP algorithms in a single device. Miniature design, low power consumption, and real-time processing requirements impose many challenges. To put many aspects of DSP in a nutshell, this final chapter has outlined principles of hearing aids. Special considerations in the design and implementation of these devices are also discussed. Although these discussions are presented from a perspective of a specific DSP application, they, to some extent, indicate some directions of future development of DSP in real-time, embedded, and system-on-chip applications.

REFERENCES

ANSI. S3.5 (1997) *Methods for Calculation of the Speech Intelligibility Index.*

ANSI S1.11-2004 (2004) *Electroacoustics—Octave-Band and Fractional-Octave-Band Filters.*

Brennan, R. and Schneider, T. (1998) "A Flexible Filter Bank Structure for Extensive Signal Manipulations in Digital Hearing Aids, in *Proceedings Of the IEEE International Symposium on Circuits Systems, CA, 1998*, pp. 569–572.

Byrne, D. and Tonnison W. (1976) "Selecting the Gain in Hearing Aids for Persons with Sensorineural Hearing Impairments," *Scandinavian Audiology*, 5, pp. 51–59.

Byrne, D. and Dillon H. (1986) "The National Acoustics Laboratories' (NAL) New Procedure for Selecting Gain and Frequency Response of a Hearing Aid," *Ear & Hearing*, 7(4), pp. 257–265.

Byrne, D., Parkinson, A., and Newall, P. (1990) "Hearing Aid Gain and Frequency Response Requirements for the Severely/Profoundly Hearing Impaired," *Ear & Hearing*, 11, pp. 40–49.

Byren, D., Dillon, H., Ching, T., Katsch, R., and Keidser, G. (2001) *NAL-NL1 Procedure for Fitting Non-Linear Hearing Aids: Characteristics and Comparisons with Other Procedures.*

Chasm, M. and Russo, R. A. (2004) "Hearing Aids and Music," *Trends in Amplification*, Vol. 8, No. 2, pp. 35–47.

Chong, K. S., Gwee, B. H., and Chang, J. S. (2006) "A 16-Channel Low-Power Non-uniform Spaced Filter Bank Core for Digital Hearing Aid," *IEEE Transactions on Circuits and Systems*, Vol. 53, No. 9, pp. 853–857.

Dillon, H., Katsch, R., Byrne, D., Ching, T., Keidser, G., and Brewer S. (1998) "The NAL-NL1 Prescription Procedure for Non-linear Hearing Aids," *National Acoustics Laboratories Research and Development, Annual Report 1997/98* (pp. 4–7). Sydney: National Acoustics Laboratories.

ISO 28961:2012, Acoustics—Statistical Distribution of Hearing Thresholds of Ontologically Normal Persons in the Age Range from 18 Years to 25 Years under Free-field Listening Conditions.

Kates, J. M. (1991) "Feedback Cancellation in Hearing Aids: Results from a Computer Simulation," *IEEE Transactions on Signal Processing*, Vol. 39, pp. 553–561.

Keidser, G. (2011) http://www.audiologyonline.com/articles/siemens-expert-series-nal-nl2-11355 (access 21/02/2016); Also Keidser G, Dillon H, Flax M, Ching T, and Brewer S. "The NAL-NL2 Prescription Procedure," *Audio Res.* 2011 May 10; 1(1): e2.

Kuo, Y-T., Lin, T-J., Li, Y-T., and Liu, C-W. (2010) "Design and Implementation of Low-Power ANSI S1.11 Filter Bank for Digital Hearing Aids," *IEEE Transactions on Circuits and Systems*, Vol. 57(7), pp. 1684–1696.

Lehto, R., Saramaki, T., and Vainio, O. (2007) "Synthesis of Narrowband Linear-Phase Filters with a Piecewise-Polynomial Impulse Response," *IEEE Transactions on Circuits and Systems*, Vol. 54, No. 10, pp. 2262–2276.

Lian, Y. and Wei, Y. (2005) "A Computationally Efficient Non Uniform FIR Digital Filter Bank for Hearing Aids," *IEEE Transactions on Circuits and Systems*, Vol. 52, No. 12, pp. 2754–2762.

Liu, C-W., Chang, K-C., Chuang, M-H., and Lin, C-H. (2013) "10-ms 18-Band Quasi-ANSI S1.11 1/3-Octave Filter Bank for Digital Hearing Aids," *IEEE Transactions on Circuits and Systems*, Vol. 60, No. 3, pp. 638–649.

Liu, Y. Z. and Lin, Z. P. (2008) "Optimal Design of Frequency Response Masking Filters with Reduced Group Delay," *IEEE Transactions on Circuits and Systems*, Vol. 55, No. 6, pp. 1560–1570.

Maxwell, J. A. and Zurek, P. M. (1995) "Reducing Acoustic Feedback in Hearing Aids," *IEEE Transactions on Speech and Audio Processing*, Vol. 3, No. 4, pp. 304–313.

Moore, B. (2004) "Dead Regions in the Cochlea: Conceptual Foundations, Diagnosis, and Clinical Applications," *Ear Hear*, 25(2), pp. 98–116.

Moore, B. C. J. (2008) "The Choice of Compression Speed in Hearing Aids: Theoretical and Practical Considerations and the Role of Individual Difference," *Trends in Amplification*, Vol. 12, No. 2, pp. 103–112.

Nielsen, L. S. and Sparso, J. (1999) "Designing Asynchronous Circuits for Low Power: An IFIR Filter Bank for a Digital Hearing Aid," in *Proceedings of the IEEE, Feb. 1999*, Vol. 87, No. 2, pp. 268–281.

Posen, M. P., Reed, C. M., and Braida, L. D. (1993) "Intelligibility of Frequency Lowered Speech Produced by a Channel Vocoder," *Journal of Rehabilitation Research and Development*, Vol. 30, pp. 26–38.

Turner, C. and Hurtig, R. (1999) "Proportional Frequency Compression of Speech for Listeners with Sensorineural Hearing Loss," *Journal of the Acoustical Society of America*, Vol. 106(2), pp. 877–886.

Index